SpringerBriefs in Astronomy

W0079802

Series Editors:

Martin Ratcliffe
Wolfgang Hillebrandt
Michael Inglis

For further volumes:
http://www.springer.com/series/10090

Tim Howard

Space Weather and Coronal Mass Ejections

 Springer

Tim Howard
Southwest Research Institute
Boulder, Colorado, USA

ISSN 2191-9100   ISSN 2191-9119 (electronic)
ISBN 978-1-4614-7974-1   ISBN 978-1-4614-7975-8 (eBook)
DOI 10.1007/978-1-4614-7975-8
Springer New York Heidelberg Dordrecht London

Library of Congress Control Number: 2013941660

© Timothy Howard 2014
This work is subject to copyright. All rights are reserved by the Publisher, whether the whole or part of
the material is concerned, specifically the rights of translation, reprinting, reuse of illustrations, recitation,
broadcasting, reproduction on microfilms or in any other physical way, and transmission or information
storage and retrieval, electronic adaptation, computer software, or by similar or dissimilar methodology
now known or hereafter developed. Exempted from this legal reservation are brief excerpts in connection
with reviews or scholarly analysis or material supplied specifically for the purpose of being entered
and executed on a computer system, for exclusive use by the purchaser of the work. Duplication of
this publication or parts thereof is permitted only under the provisions of the Copyright Law of the
Publisher's location, in its current version, and permission for use must always be obtained from Springer.
Permissions for use may be obtained through RightsLink at the Copyright Clearance Center. Violations
are liable to prosecution under the respective Copyright Law.
The use of general descriptive names, registered names, trademarks, service marks, etc. in this publication
does not imply, even in the absence of a specific statement, that such names are exempt from the relevant
protective laws and regulations and therefore free for general use.
While the advice and information in this book are believed to be true and accurate at the date of
publication, neither the authors nor the editors nor the publisher can accept any legal responsibility for
any errors or omissions that may be made. The publisher makes no warranty, express or implied, with
respect to the material contained herein.

Printed on acid-free paper

Springer is part of Springer Science+Business Media (www.springer.com)

Preface

The term "Space Weather" has reached a greater sense of public awareness in recent years. When used in the press, it usually refers to a large disturbance of the Earth's magnetic field called a (geo)magnetic storm, but the term actually applies to any phenomenon in the solar system that is caused by solar activity. The Sun is only now coming out of an extended period of inactivity and so when the chances of increased space weather effects at the Earth increase even slightly it is immediately reported in the press. Unfortunately most of these have proven to result in nothing significant at the Earth. The Dst index, the universal indicator of geomagnetic activity, has only dropped below $-100\,$nT six times, and has not once dropped below $-150\,$nT, since December 2006. In other words, so far in the current solar cycle there has been no geomagnetic activity that could be regarded as severe.

So why all the press interest in the possibility of severe space weather? Partly, it's because of its rarity in the current solar cycle. This may change in the next year or so as we reach the maximum in solar activity (what we call solar maximum), and it is worth noting that during the last solar cycle the most severe space weather events were caused in the declining stage of the cycle as the solar activity ramped back down towards solar minimum.

The other, and most important reason for the increased interest in space weather is because of the potential damaging effects that can result from it. We live in a technological age with nanotechnology and electromagnetic communication, and as we become more advanced our technology becomes smaller and more sophisticated. Electronic devices carry smaller electrical circuits, and so even minor variations in the magnetic environment could induce damaging currents, potentially short-circuiting or overloading these electric systems. Sudden magnetic variations can disrupt electromagnetic transmission, potentially damaging satellite and ground communications. Also are increased radiation effects in the high-altitude atmosphere, where increasing numbers of people travel on intercontinental flights. These hazards are already known to be potentially serious, and they will become more significant as our technology and traveling habits become more sophisticated.

Because of these hazards governments, the private sector, and now the general public are becoming more aware of space weather, and to how we can predict and mitigate its most severe cases.

What causes space weather? Or more precisely, what causes (geo)magnetic storms? Read any news report on the last predicted magnetic storm and you will probably find that they are caused by solar flares. This is not true, or rather not true most of the time. Flares are sources of great concentrations of energetic particles, and those particles can be magnetically guided to the Earth thereby causing a magnetic storm. Solar flares, however, are highly localized phenomena, and even if the particles are able to "fan-out" from their source regions, the Earth would need to well placed at the right time to receive the bulk of energetic particles arriving from a solar flare. While it would not be correct to say that this never happens, it is not what causes most magnetic storms. The real culprit is the coronal mass ejection.

Some of the better news outlets are beginning to mention coronal mass ejections (CMEs) more regularly when reporting on space weather, and while the phenomenon is gaining recognition it is still widely believed that solar flares cause magnetic storms. A CME is a large eruption of plasma and magnetic field from the Sun. It is much larger in size, and carries much more energy than the flare, and its interaction with the Earth's magnetic field is responsible for major space weather. It is also a physical mechanism for the removal of magnetic complexity and built-up magnetic energy from the Sun, and therefore plays an important role in solar evolution.

This Brief introduces the coronal mass ejection as the phenomenon responsible for magnetic storms at the Earth. We discuss the physics behind how CMEs interact with the Earth and the processes by which they build up energy at the Sun and erupt. We explore the history of CME observation, how we observe and track them, the physics behind their onset and evolution, and how they cause space weather.

The author published a larger and more detailed introduction to CMEs last year, called *Coronal Mass Ejections: An Introduction* (Springer, 2011). This Brief is mostly an abridged and updated version of that larger text, with an additional chapter (Chap. 5) dedicated to the latest developments that have occurred since the publication of that book. I have gone to great lengths, for example, to significantly reduce the mathematical content in this Brief, but a small amount does remain as in such a complex topic as CMEs it is unavoidable.

The purpose of this Brief is to introduce the concept of the CME to a non physics audience: to science students, scientists in fields other than space physics, or to the general public seeking to find out more about space weather. If there is a single message the reader should take away from this book it is this: The most severe space weather effects, (geo)magnetic storms, are not caused by solar flares. They are caused by coronal mass ejections. Therefore if one is interested in studying the causes of severe space weather it is the CME, and not the flare, that we should be investing our resources into. As this Brief shows, there is much that is not yet understood about this important and fascinating phenomenon.

Boulder, Colorado, USA   Tim Howard

Contents

Acronyms

3-D	Three dimensions, or three-dimensional.
ACE	*Advanced Composition Explorer*: Spacecraft launched in 1997 into L1 orbit, designed to monitor the interplanetary medium with a suite of in-situ instruments.
AIA	Atmospheric Imaging Assembly: White light, UV and EUV imaging suite on board *SDO*.
A_p	An index describing geomagnetic activity. It is a daily index derived from the eight three-hourly a indices. A_p may be regarded as a daily average of the three-hourly K_p index.
AGU	American Geophysical Union.
AU	Astronomical Unit: The average distance of the Earth from the Sun ($\sim 1.5 \times 10^8$ km).
CCMC	Community Coordinated Modeling Center: NASA-based site for space weather related models (ccmc.gsfc.nasa.gov).
CDS	Coronal Diagnostic Spectrometer: EUV spectrometric imager on board *SOHO*.
CELIAS	Charge, ELement, and Isotope Analysis System: Particle detector on board *SOHO*.
CIR	Corotating Interaction Region: Interaction in the heliosphere between fast and slow solar wind streams.
CME	Coronal Mass Ejection: An eruption of plasma and magnetic field from the Sun.
COR	Coronagraphs on board *STEREO*: There are two, COR1 (inner) and COR2 (outer).
COSTEP	COmprehensive SupraThermal and Energetic Particle analyser: Particle detector on board *SOHO*.
C/P	Coronagraph/Polarimeter: White light coronagraph/polarimeter on board *SMM*.

Dst   An index describing geomagnetic activity, derived from mea-
                 surements of the equatorial region of the geomagnetic field, and
                 thus monitors the Earth's ring current. When the ring current is
                 enhanced, the equatorial geomagnetic field is reduced, indicated
                 by a reduction in the Dst index. A sudden large decrease in the
                 Dst index is an indicator of a geomagnetic storm.

EIS   Extreme-ultra-violet Imaging Spectrometer: On board *Hinode*.

EIT   Extreme-ultra-violet Imaging Telescope: EUV imager on board
                 SOHO.

ERNE   Energetic and Relativistic Nuclei and Electron experiment: Parti-
                 cle detector on board *SOHO*.

ESA   European Space Agency: Main institution of European space
                 research and exploration.

EUV   Extreme Ultra-Violet: The highest frequency band of the ultra-
                 violet spectrum (prior to the X-ray band).

EUVI   Extreme-Ultra-Violet Imager: On board *STEREO*.

EVE   Extreme-ultra-violet Variability Experiment: EUV irradience in-
                 strument on board *SDO*.

FOV   Field of view: An area observed by an imager.

GLE   Ground Level Enhancement: An increase of cosmic ray intensity
                 at the Earth, resulting from the arrival of solar energetic particles.

GOLF   Global Oscillations at Low Frequencies: Helioseismology instru-
                 ment on board *SOHO*.

GSFC   Goddard Space Flight Center: The main research wing of NASA.

Hα The hydrogen emission line in the visible light spectrum denoting
                 the α transition: A common wavelength at which visible light
                 telescopes observe the Sun.

HAE   Helium Abundance Enhancement: A sudden increase in the abun-
                 dance of helium in the interplanetary medium, typically following
                 an interplanetary shock.

HAF   Hakamada-Akasofu-Fry model (Also known as HAFv2): Model
                 describing the propagation of an interplanetary shock.

HAO   High Altitude Observatory: Based in Boulder, CO.

HEOS-2   *Highly Eccentric Orbit Satellite 2*: A European spacecraft launched
                 in 1972 designed to study the high altitude magnetosphere and
                 near-Earth interplanetary medium.

HESSI   The original name of the *RHESSI* spacecraft.

HI   Heliospheric Imager: White light heliospheric imagers on board
                 STEREO. There are two, HI-1 (inner) and HI-2 (outer).

Hi-C   *High resolution Coronal imager*. NASA sounding rocket mission
                 launched in July 2012.

ICE   *International Cometry Explorer*: Originally the *ISEE-3* spacecraft
                 but left its L1 orbit in 1982 to investigate Comet P/Giacobini-
                 Zinner.

ICME	Interplanetary Coronal Mass Ejection: The old term for the heliospheric counterpart of a CME.
IMF	Interplanetary Magnetic Field: The magnetic field within the interplanetary medium.
IMP-8	*Interplanetary Monitoring Platform 8*: A spacecraft launched into High Earth Orbit in 1973, designed to monitor the interplanetary medium.
IPS	InterPlanetary Scintillation: A short-duration distortion of the signal from a distant radio source as a result of a dense structure passing between it and the observer.
IR	Infra-Red: The region of the electromagnetic spectrum in the frequency band immediately smaller than visible light.
IRIS	Interface Region Imaging Spectrograph. Solar imaging space-craft. Scheduled for launch in December 2012.
ISEE-3	*International Sun-Earth Explorer 3*: A spacecraft launched in 1978 into the L1 Lagrange point, where it monitored the interplanetary medium before later becoming *ICE* and heading off to explore a comet.
ISPM	Interplanetary Shock Propagation Model: Model describing the propagation of an interplanetary shock.
JPL	Jet Propulsion Laboratory: NASA scientific and hardware group based at Caltech.
K_p	An index describing geomagnetic activity. K_p is determined every 3 h and ranges from 0 (quiet) to 9 (active) in increments of $\frac{1}{3}$. It is derived using a network of ground-based magnetometers measuring the most disturbed horizontal component.
L1	The first Lagrange point: A location on the Sun-Earth line where the gravitational pull of the Sun is exactly canceled out by that of the Earth. This is around 1.5×10^6 km from the Earth, or 1 % of the distance from the Earth to the Sun.
LASCO	Large Angle Spectroscopic COronagraph: Coronagraphs on board *SOHO*. There were three, C1 (inner), C2 (middle) and C3 (outer) but C1 ceased to operate following the *SOHO* incident in 1998.
LOS	Line Of Sight: The vector from the observer through the point of interest and out to infinity.
MDI	Michelson Doppler Imager: Helioseismology instrument on board *SOHO*.
MESSENGER	*MErcury Surface, Space ENvironment, GEochemistry and Ranging*. Spacecraft currently in orbit around Mercury.
MHD	Magnetohydrodynamics.
MIR	Merged Interaction Region: Regions of compressed plasma and dense magnetic field at very large distances from the Sun, often caused as a result of a CME interacting with the surrounding medium.

NASA	National Aeronautic and Space Administration: The leading authority in US space research and travel.
NOAA	National Oceanic and Atmospheric Administration.
NRL	Naval Research Laboratory.
OSO-7	*Orbiting Solar Observatory 7*: A solar observing spacecraft launched in 1971.
P78-1	A spacecraft launched in 1979 which carried the *Solwind* coronagraph.
$R_\odot$	Unit of distance: 1 $R_\odot$ = 1 Solar radius $\sim$695,500 km.
R_E	Unit of distance: 1 R_E = 1 Earth radius $\sim$6,360 km.
R_H	Unit of distance: 1 R_H = 1 Hermian (Mercury) radius $\sim$2,440 km.
R_J	Unit of distance: 1 R_J = 1 Jovian (Jupiter) radius $\sim$71,500 km.
R_S	Unit of distance: 1 R_S = 1 Saturn radius $\sim$60,270 km.
R_U	Unit of distance: 1 R_U = 1 Uranus radius $\sim$25,560 km.
R_N	Unit of distance: 1 R_N = 1 Neptune radius $\sim$24,770 km.
RAISE	*Rapid Acquisition Imaging SpectroscopE*. A solar-observing spectroscope launched on a sounding rocket in August 2010.
RHESSI	*Ramaty High Energy Solar Spectroscopic Imager*: Spacecraft launched in 2002 designed to monitor the Sun with an x-ray imager and spectrometer.
S(S)C	Sudden (Storm) Commencement: The onset of a geomagnetic storm following an abrupt increase in the strength of the horizontal component of the geomagnetic field (known as a sudden impulse).
SDO	*Solar Dynamics Observatory*: Spacecraft launched in February 2010 designed to monitor the Sun with a suite of imagers.
SECCHI	Sun Earth Connection Coronal and Heliospheric Investigation: Imaging instrument suite on board *STEREO*.
SEP	Solar Energetic Particle: High-energy particles originating from the Sun and observed in the heliosphere.
SMEI	Solar Mass Ejection Imager: White light heliospheric imager on board *Coriolis*.
SMM	*Solar Maximum Mission*: A spacecraft launched in 1980 designed to measure the Sun with a suite of instruments and spectrometers.
SOHO	*SOlar and Heliospheric Observatory*: A spacecraft launched in 1995 into the L1 point with a suite of solar observing instruments on board.
STEREO	*Solar TErrestrial RElations Observatory*: Pair of solar-observing spacecraft sharing an orbit about the Sun with the Earth.
STEREO-A	The *STEREO* spacecraft that is leading the Earth: All instruments on board this spacecraft also have the suffix "-A" (e.g. COR2-A, HI-1A).
STEREO-B	The *STEREO* spacecraft that is following the Earth: All instruments on board this spacecraft also have the suffix "-B" (e.g. COR1-B, HI-2B).

STOA	Shock Time Of Arrival: Model describing the propagation of an interplanetary shock.
SUMER	Solar Ultra-violet Measurements of Emitted Radiation: UV spectrometer on board *SOHO*.
SUMI	*Solar Ultraviolet Magnetograph Investigation*. NASA sounding rocket mission launched in July 2012.
SWAN	Solar Wind Anisotropies: Neutral particle detector on board *SOHO*.
SWPC	Space Weather Prediction Center: NOAA-based prediction center for space weather.
TRACE	*Transition Region And Coronal Explorer*: Spacecraft launched in 1998 designed to monitor the Sun with a UV/EUV imager.
UV	Ultra-Violet: The region of electromagnetic spectrum in the frequency band immediately larger than visible light.
UVCS	Ultra-Violet Coronagraph Spectrometer: UV spectrometer on board *SOHO*.
VEX	*VEnus eXpress*: Spacecraft in orbit around Venus.
VIRGO	Variability of Solar Irradiance and Gravity Oscillations: Helioseismology instrument on board *SOHO*.
WSA	Wang-Sheeley-Arge model: Photospheric and coronal field and plasma model based on photospheric measurements.

Chapter 1
Introduction

Consider the following events in ancient and recent history:

- **34AD:** Roman Emperor Tiberius dispatches troops to investigate a suspected fire at the port of Ostia.
- **1724AD:** George Graham in England and Anders Celsius in Sweden simultaneously find a deviation in their compass needle that lasts more than a day [17].
- **September 1859:** Telegraph systems across Europe and North America simultaneously fail [6].
- **September 1941:** Television and radio broadcasts are interrupted, including 15 min of a baseball game between the Brooklyn Dodgers and the Pittsburgh Pirates. The Pirates scored 4 runs from a 0-0 drawing game during those missing minutes, causing great frustration for the many viewers and listeners [26].
- **July 1979:** The first US space station *Skylab* re-enters Earth's atmosphere ahead of schedule, crashing into the Australian desert.
- **March 1989:** Power stations fail across Canada and are damaged or disrupted in the UK, Scandinavia and the US. At the same time, mysterious readings appear in the orbiting Space Shuttle *Discovery*'s sensors [26].
- **January 1994:** Three spacecraft: *Anik E1*, *Anik E2* and *Intelsat-K*; fail at the same time [1].
- **October 2003:** Power stations fail across North America and are damaged across Europe, along with one Japanese spacecraft (*ADEO-II*). Other spacecraft, including the Japanese *Hayabusa* and *Mars Odyssey* are damaged [1].

What do these events have in common? Here's a clue. At all of the times listed above the aurora was observed at locations on the planet where it is very rarely seen: at low latitudes. It was, in fact, the aurora that was observed by Tiberius who mistook its red glow for a fire. It was unexpected there because Ostia lay just south of 42 °N. Was it the aurora that caused all that damage listed above? Not entirely, but it was an excellent indicator of what was occurring during those times. The real question to ask is what was causing the aurora to occur at low latitudes in the first place?

T. Howard, *Space Weather and Coronal Mass Ejections*, SpringerBriefs in Astronomy, DOI 10.1007/978-1-4614-7975-8_1, © Timothy Howard 2014

1.1 Magnetic Storms

What was occuring during the times listed above is the most extreme form of space weather. "Space Weather" is a collective term used to describe environmental behavior at Earth and other planets that arise from activity in space (i.e., from outside the planetary environment). All of this activity originates at the Sun in some form. The most extreme forms of space weather are magnetic storms. A magnetic storm, a term coined by von Humboldt in 1805 (also called a geomagnetic storm when occurring at Earth), is a large disruption of a planet's magnetic field, brought about by the injection of great quantities of energy and energetic particles from space into its magnetosphere. One consequence of this is the opening up of typically closed (geo)magnetic field lines, exposing a larger region of the atmosphere to the space environment than is normal. The physics of how this occurs is discussed later in this book (Sects. 1.6 and 4.3), but this is why the aurora is observed at lower latitudes during times of increased magnetic activity.

It was this enhanced geomagnetic activity that was observed by Graham and Celsius in 1724, and the highly variant magnetic fields can induce large unexpected currents in long electrical wires, such as those transmitting telegraph signals and are connected with power stations. This is what caused the telegraphs to fail in 1859 and what probably caused the failure of the power stations. Spacecraft, which orbit in the Earth's upper atmosphere where these effects are strong, can be damaged from the increased bombardment of energetic particles, from short-circuiting from induced currents in their circuit boards, or from increased exposure to damaging radiation. Spacecraft can even be brought down prematurely from increased atmospheric drag brought about by a larger concentration of particles in their environment; such was the case with *Skylab* in 1979. Along with these damaging effects to technology; during magnetic storms radiation that is harmful to humans is increased, especially at high altitudes and high latitudes, posing a radiation hazard for airline passengers and crew, and for astronauts. It is now routine for airline flights to be re-directed to lower latitudes when increased geomagnetic activity is predicted. All of these problems will continue to be enhanced as our technological abilities improve and our electrical devices become smaller and more sophisticated.

1.2 Coronal Mass Ejections

What causes magnetic storms? Any phenomena that could dump great quantities of energetic particles in a short time could do it, but during a typical magnetic storm at the Earth over 6 GW of power are dumped into the magnetosphere [18]. There are really only two phenomena that can provide the particle concentrations and energies required to deliver such vast quantities of power globally across the Earth. These are energetic particles from solar flares, and coronal mass ejections.

Solar flares are the most commonly recognized eruptive phenomena from the Sun. They are seen as bright emissions at the Sun and have been observed since the nineteenth century. They eject large quantities of highly energetic particles into space, and those particles can (and often do) collide with the Earth, other planets, and spacecraft. While they do cause magnetic storms (the March 1989 storm was begun with the arrival of energetic particles from a flare [26]), they are not the cause of most of them, largely because solar flares are highly localized phenomena at the Sun. A flare needs to be in just the right place on the solar surface for its energetic particles to be connected with the Earth. The more common cause of magnetic storms at Earth are coronal mass ejections.

1.3 Overview

Given that it is the coronal mass ejection (CME) that is the primary cause of severe space weather at the Earth, an understanding of this phenomenon is now essential to any who work in the technological field, particularly those working at high-altitudes or with space technology. A complete understanding CMEs and their effects on space weather requires not only an understanding of solar physics, but also the physics of the interplanetary medium, and the Earth's magnetosphere and ionosphere. It is very rare for research groups to bridge the gaps between these highly specialized subjects.

Last year (2011), Springer published an introductory text to coronal mass ejections written by the author [19]. This was intended as a one-stop introduction to CMEs, presented in a non-specialized way, and provided an exhaustive list of references bringing us to the status quo regarding CME study. The intended audiences were (post)graduate students of physics or mathematics, or scientists in other areas who would like a general introduction to the subject. The text was somewhat heavy on theory in places, and parts would make heavy reading for those not well-versed in applied mathematics. It is the intention of this Brief to introduce coronal mass ejections with a lighter theoretical touch; although some theory is still here (this is inevitable), it is not with the level of depth and detail as that discussed in the 2011 book. The intended audience remains those in the scientific research community, but it is also intended to be of value to high school and college teachers and to undergraduate students as well. This is part of the Springer Briefs series, and therefore provides a brief introduction to CMEs: what they are, how we observe them, and why they are important. The emphasis here is on the importance of CMEs to space weather, but we also briefly discuss their solar origins. We bring the reader up to date on the latest developments in CME study, as research into them is ongoing and ever-evolving.

The outline of this Brief is as follows. In this first chapter we introduce the concept of the coronal mass ejection and outline what they are, how and why they are formed, how they are observed, and why they are important to space weather. The history of CME observation is summarized in Chap. 2, beginning with observations of the corona and of space weather, to the detailed solar studies in the mid-nineteenth century, and through the space age. In Chap. 3 we discuss the means by which CMEs are observed and modeled and Chap. 4 discusses their relevance to space weather. We conclude with a discussion of the latest developments in CME study in Chap. 5.

1.4   What Is a CME?

A coronal mass ejection (CME) is a large eruption of plasma and magnetic field from the Sun. It can contain a mass larger than 10^{13} kg [25] and may achieve a speed of over a 1,000 km/s [22]. A typical CME has a mass of around 10^{11}–10^{12} kg and has a speed between 300 and 1,000 km/s. It also typically spans several 10's of degrees of heliographic latitude (and probably longitude). By comparison, the Earth has a mass of around 6×10^{24} kg and is around $(5 \times 10^{-3})^{\circ}$ in heliographic latitude. Images of a CME are shown in Fig. 1.1. These are images taken at around the same time from three different viewpoints, two (red and blue) from the COR2 coronagraphs on the two *STEREO* spacecraft, and one from the LASCO coronagraph on board *SOHO*. LASCO (orange) provides a view of how the CME looked when viewing it from the Earth, and the *STEREO* spacecraft are observing it at angles of around 70° on either side of the line from the Sun through the Earth. These images not only show the detailed structure comprising the CME, but also how its appearance can change significantly depending on the location from which it is viewed.

CMEs may erupt from any region of the corona but are more often associated with lower solar latitude regions, particularly near solar minimum (the minimum of solar activity throughout the 11 year solar cycle). They are also often associated with the heliospheric current sheet (the surface where the polarity of the Sun's magnetic field changes) [21], but this by no means explains the location of every CME. They erupt from the Sun with an occurrence of around once a day during typical solar minimum, and around four or five times per day during solar maximum [31]. Only a small percentage of CMEs are directed toward the Earth.

CMEs erupt through the solar corona, out into the solar wind, and propagate through interplanetary space. They sweep past the planets and into the heliosphere and they have been measured as far out as the termination shock [27]. They carry with them the complex magnetic fields brought from the Sun, along with solar material at their source, and they also sweep up solar wind material as they travel. By the time they reach distances of around 1 AU from the Sun, their magnetic properties and ionic composition may be considerably different to their original states. It is known, for example, that the internal magnetic structure comprising the CME flux rope (called a magnetic cloud), can be distorted in structure as it propagates through the solar wind [20] and that filament material, which is often the brightest component of the CME when observed in the corona (see Fig. 1.1), is rarely observed near 1 AU [5, 23].

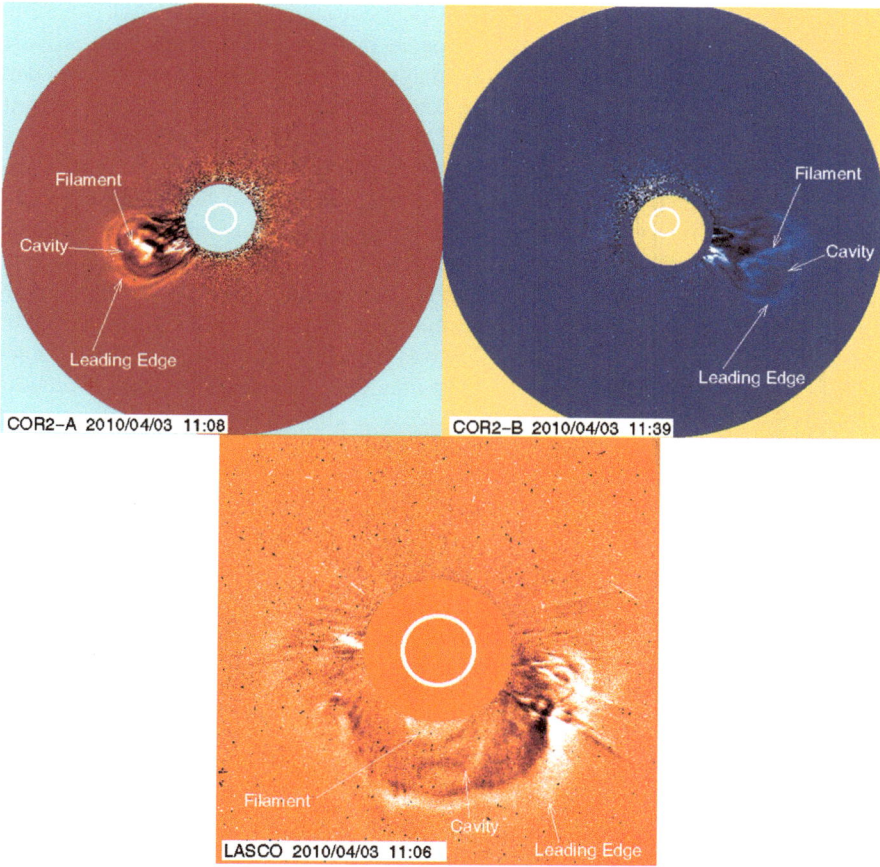

Fig. 1.1 A coronal mass ejection (CME) as observed from coronagraphs from three separate locations. The *top left panel* (*red*) shows it as seen by *STEREO-A*/COR2 when it was around 67° ahead Sun-Earth line; the *top right panel* (*blue*) is from *STEREO-B*/COR2 at around 71° behind the Sun-Earth line; and the *lower middle panel* (*orange*) is from *SOHO*/LASCO-C2 located roughly on the Sun-Earth line. Each image was taken on the same day (3 April 2010) and around the same time (11:08UT for COR2-A, 11:39UT for COR2-B, and 11:06UT for LASCO). These images show the main components that are common for most CMEs (labeled) and the different structures demonstrate how the appearance can change drastically depending on the location of the observer

Up until the last decade, CMEs were observed exclusively in two regimes: Close to the Sun by coronagraphs and other imagers that observe the solar corona; and in-situ by spacecraft near the Earth that measure their properties as they pass through a traveling CME.[1] This left a large observing gap between around 0.1 AU (the

---

[1] We leave aside for the moment those CME detections made using interplanetary scintillation and those spacecraft not near the Earth, that we discuss in Chap. 2.

outer edge of the field of view of the widest coronagraphs) and 1 AU, where the
majority of in-situ spacecraft orbit. Imagers that observe near the Sun provided
large-scale images of the CME with no magnetic field or ionic compositional data,
and in-situ measurements provided the latter measurements in great detail but with
no context for the large-scale structure of the CME. It was therefore difficult to
identify exactly which components measured in-situ near 1 AU arose from which
components observed by coronagraphs near the Sun. The two classes of observation
were therefore regarded almost as separate measurements, and the phenomena
measured by each was also termed differently. The erupting structure seen near the
Sun (Fig. 1.1) was called a CME, and the in-situ signature, known to have originated
as the CME near the Sun, was called an interplanetary CME or ICME [10]. In 2003,
the Solar Mass Ejection Imager (SMEI) was launched. SMEI was the first of a new
class of instrument, called a heliospheric imager. Heliospheric imagers (Sects. 2.6,
3.1.2 and 3.1.3.3) are able to observe CMEs at all stages of their evolution from
the Sun to 1 AU and so there is no longer a gap between observations. Because of
this, the term ICME has since been dropped by many workers and the connection
between structural components observed in coronagraphs and measured in-situ is
now being made [9, 20].

1.5   The Structure and Composition of CMEs

During onset and through the early eruption stages, the physics governing the CME
are almost entirely magnetic [7, 14]. The actual circumstances around which CMEs
erupt are unknown but the vast majority of models involving CME onset describe
the evolution of magnetic fields (Sect. 3.2.1 discusses some of these models). The
magnetic structures surrounding the CME near the Sun are generally regarded as
comprising a launching central "core" magnetic field, and other surrounding coronal
fields that are either locally closed at the Sun or locally open with one end extending
into the solar wind. The circumstances behind the launch of the core and what
happens to the surrounding field are unknown.

Upon inspection of the left (red) panel of Fig. 1.1 we can imagine the CME as
a kind of light bulb structure, with a leading edge, followed by a central cavity,
followed by a bright filament. This is referred in the literature as a "classic" three-
part CME [23]. In terms of magnetic fields, it is now believed that the erupting CME
magnetic core is actually the cavity component of this three-part structure [15, 20],
the trailing bright filament is cooler filament (prominence) material arising from
lower in the solar atmosphere (often observed as a dark filament in $H\alpha$ solar disk
images that disappears), and the leading bright front is coronal and solar wind
material swept up ahead of the erupting CME core field. Figure 5.4b shows a cross-
sectional view of the evolution of the different magnetic components of a CME.

At large distances from the Sun, where a single track through the CME is
measured by non imaging in-situ instruments, there is also commonly a three-part
structure observed, but they are not the same three parts that are observed by

coronagraphs. In a typical (uncomplicated) CME structure we measure the central core field, that often has a highly structured spiral field configuration known as a magnetic cloud (Sect. 2.3.1), which also usually, but not always, has a lower density than its surroundings. We also measure the bright leading edge as a turbulent region preceding the magnetic cloud which is called the sheath region. Ahead of that, especially in cases of fast CMEs, a forward shock is often detected. Forward shocks are easily identified by a sharp increase in magnetic field strength and solar wind density and speed but also, since in-situ measurements provide precise measurements of these parameters, they can be tested as to whether they obey the Rankine-Hugoniot relations concerning mass, momentum and energy conservation across shock fronts (see, e.g., [13]). Whether there is in fact a forward shock ahead of the bright leading edge in the coronagraph images remains unknown, but is likely. The filament component, mentioned previously, is not typically apparent in in-situ measurements. Figure 1.2 shows a series of in-situ signatures of a well-defined CME.

1.5.1 Ionic Composition

We currently have no way of identifying the ionic composition of CMEs when they are near the Sun. This is because the instruments that observe them there, coronagraphs, do not contain any spectral information from which we would typically extract such data (van Houten [29] discusses why useful spectral information is lost in the mid and high corona). We can, and regularly do, measure CME-related signatures in the low corona using spectrographs (*SOHO*/CDS, *Hinode*/EIS and *SDO*/EVE are examples of such instruments).[2] The extent to which these associated phenomena comprise the CME as observed by coronagraphs remains unknown. The Solar Probe Plus mission, scheduled for launch in 2018, is intended to pass within nine solar radii of the Sun (well within the fields of view of many coronagraphs), and will provide direct measurements of these properties.

Far from the Sun, we routinely measure the internal properties of CMEs using in-situ spacecraft, and have done so now for decades (see Sect. 2.2.1). Direct measurements of CMEs from the early 1970s revealed a helium abundance enhancement following interplanetary forward shocks, and high ionization states of oxygen and iron [2–4, 12]. CMEs are also known to contain heavy elements in high ionization states, such as Fe^{10+} and even Fe^{16+} [16, 24]. They also sometimes contain cooler ions, such as singly-charged helium, magnesium and neon [4]. The high-temperature ions are generally regarded to originate low in the solar corona or from heating during the launch of the CME itself, while the low temperature ions, as

---

[2]While we will spend little time discussing these phenomena that are related to CMEs in this Brief, a discussion of them appears in Chap. 7 of the author's introductory text on CMEs [19].

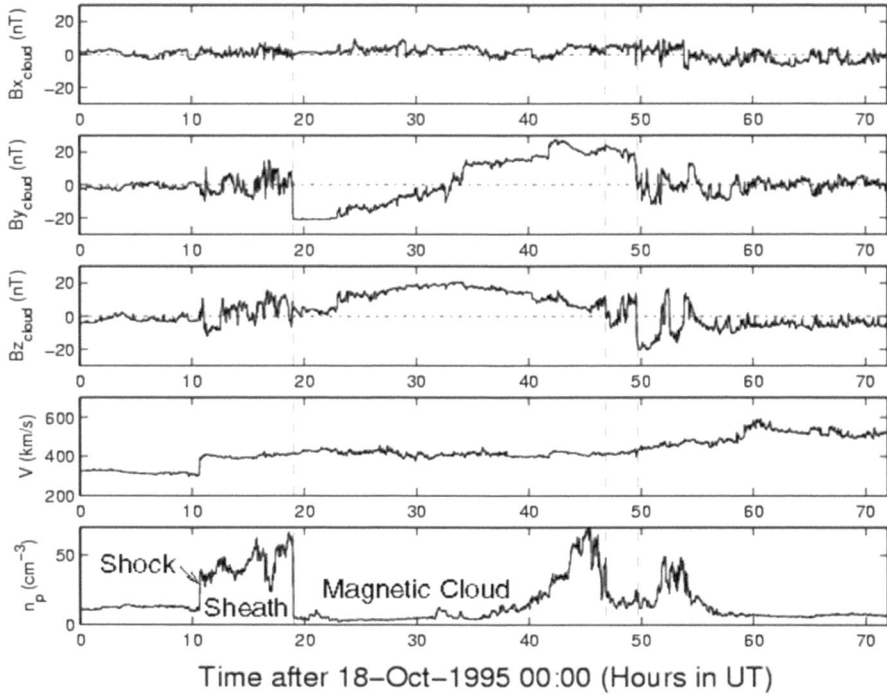

Fig. 1.2 *Wind* measurements of the magnetic cloud observed on 18–20 October 1995 (from Fig. 1 of Dasso et al. [8]). From the *upper* to the *lower panel*: radial, azimuthal, and axial components of the magnetic field, bulk velocity, and proton plasma density, as a function of time (in hours, after 00:00UT, October 18 1995). *Dashed lines* mark the front boundary of the cloud (18:58UT, October 18 1995). The possible ends of the cloud (October 19, 22:54UT and October 20, 01:38UT) are indicated with the *dash-dotted lines*

mentioned before, are probably associated with the filament material [5, 23]. While different CMEs have different compositions there do appear to be some repeatable patterns, like He^{++}/H^{+} and Fe^{16+} enhancements, that are common to many CMEs.

1.6   Why Do CMEs Erupt?

The most logical (and likely) explanation for why CMEs erupt is that they are processes by which the Sun removes vast quantities of built-up magnetic complexity and energy [7, 14]. Although the Sun to the naked eye appears as a smooth object, when observed up-close and at wavelengths other than the visible, the appearance is far from smooth. Through the upper third of the solar atmosphere, ending at the edge of the photosphere, energy is transmitted primarily through convection (hence the term "convection zone"). Another example of convection is boiling water on a stove. At the Sun, just as on the stove, convection cells are established, and the appearance of the top of the convection zone is one of chaotic migration of these cells about the solar photosphere. These cells are comprised magnetic fields, which are forced into such cells and migrate around the Sun. A picture emerges of a highly complex magnetic field, where existing fields above the surface of the Sun interact with emerging fields from below, and the whole system is in a state of constant change. Magnetic flux or complexity are manifest on the Sun as active regions, sunspots, filaments or arcades. The corona is in a state of constant expansion, and it is through this process that much of the energy and field is removed (this expansion forms the solar wind.) Sometimes, however, higher up in the solar atmosphere, a stable closed magnetic configuration can evolve which prevents the local corona from expanding into the solar wind. The field will increase in complexity and the stored energy will continue to be built-up, as new magnetic flux emerges from beneath. The stored energy cannot build up indefinitely and so eventually this system must be released. It does so explosively, in the form of a CME. CMEs have been known to achieve over 10^{39} J in kinetic energy alone [25]. Smaller energy releases, such as those from solar flares, also provide a means by which the Sun can remove stored energy. A strong solar flare releases around 10% of the energy carried by a typical CME [11, 30].

Section 3.2.1 discusses the onset mechanisms and the problems associated with launching CMEs in more detail. While the exact launch process is not known, it seems most likely that the energy released originates as stored magnetic energy and that the explosive release of that energy is likely because of a pre-existing stable magnetic structure impeding the expansion of the corona.

1.7 How Do CMEs Contribute to Space Weather at the Earth?

We have already in this chapter discussed the deleterious effects of severe space weather events (magnetic storms) at the Earth. We have also stated that coronal mass ejections are responsible for the majority of these storms, although energetic particles from solar flares are sometimes responsible too. Very severe storms can actually arise from a "double punch", where energetic particles from the solar flare arrive within minutes of the launch at the Sun (solar energetic particles travel at relativistic speeds) and then the CME arrives at the Earth a day or so later. While the direct injection of energetic particles arriving from the solar wind can be imagined as being fairly straightforward, the process is more complicated with CMEs as they are magnetic structures that are closed at the Sun and the geomagnetic field is closed at the Earth. So how can the magnetic field of the CME interact with that of the Earth?

Details of how CMEs deliver energetic particles and energy to the Earth's magnetosphere are provided in Chap. 4. The key lies in a process known as magnetic reconnection (Sect. 4.3).

The Earth is enclosed within the solar wind, which is a continually-flowing "ocean" of plasma and magnetic field moving outward from the Sun. For the most part, the solar wind is deflected around the Earth by the geomagnetic field, which itself is distorted by this interaction (see Figs. 1.3 and 4.1). The dynamics of this distortion and the interaction with the Earth's atmosphere result in a buildup of plasma in various regions around the Earth (Fig. 4.2). This resulting combination of plasma and field is called the magnetosphere, and its behavior is strongly influenced by that of the surrounding solar wind. The aurora, for example, is caused by particle precipitation from the solar wind and accumulated plasma in the magnetosphere, from particles that can directly enter the Earth's atmosphere via the divergence of magnetic field lines near the poles (the so-called cusp region), and the open field lines towards the anti-Sunward side of the magnetosphere. These energetic particles provide energy to the particles of the atmosphere, resulting in an emission of light.

The direction of the geomagnetic field is consistently northward, that is, it flows from south to north (geomagnetic north is in the same hemisphere as geographic south). When a strong southward-directed magnetic field arrives at the magnetosphere, that field can undergo magnetic reconnection with the geomagnetic field, allowing the two fields to temporarily connect. This causes more field lines on the dayside of the magnetosphere to open, resulting in a movement of the cusp (the part open to the solar wind) toward the equator (hence the aurora moves towards the equator). Figure 4.3 illustrates this phenomenon. This figure also shows that a northward-directed interplanetary magnetic field will not reconnect with the magnetosphere and will therefore have minimal effect on geomagnetic activity. It should be noted that even if this were to occur, if the solar wind arrival was sufficiently faster and/or denser than its environment, then the ram pressure alone may cause a disruption to the magnetosphere. It should also be noted that magnetic reconnection still does occur on the nightside of the magnetosphere with a northward interplanetary magnetic field, but this does not normally lead to storm conditions.

This simple example demonstrates the dependence of not only the structure and dynamics of the magnetosphere on the behavior of the solar wind, but also a means by which energy and energetic particles may enter the Earth's magnetosphere. A CME may be regarded as an extreme version of the solar wind, as it contains high-speed plasma and magnetic field just as the solar wind does, only it is greatly enhanced. If the internal magnetic field of the CME is predominantly southward, two major processes may occur:

1. Magnetic reconnection exposes the Earth to the plasma contained within the CME, which is injected directly into the geomagnetic field. Reconnection causes closed field lines to open, exposing them to the solar wind and allowing a larger proportion of the Earth's atmosphere to be injected with its plasma.
2. The increased pressure impacting the magnetosphere causes it to compress and closed magnetic field lines to be reduced in size. This results in a further expansion of the auroral ovals.

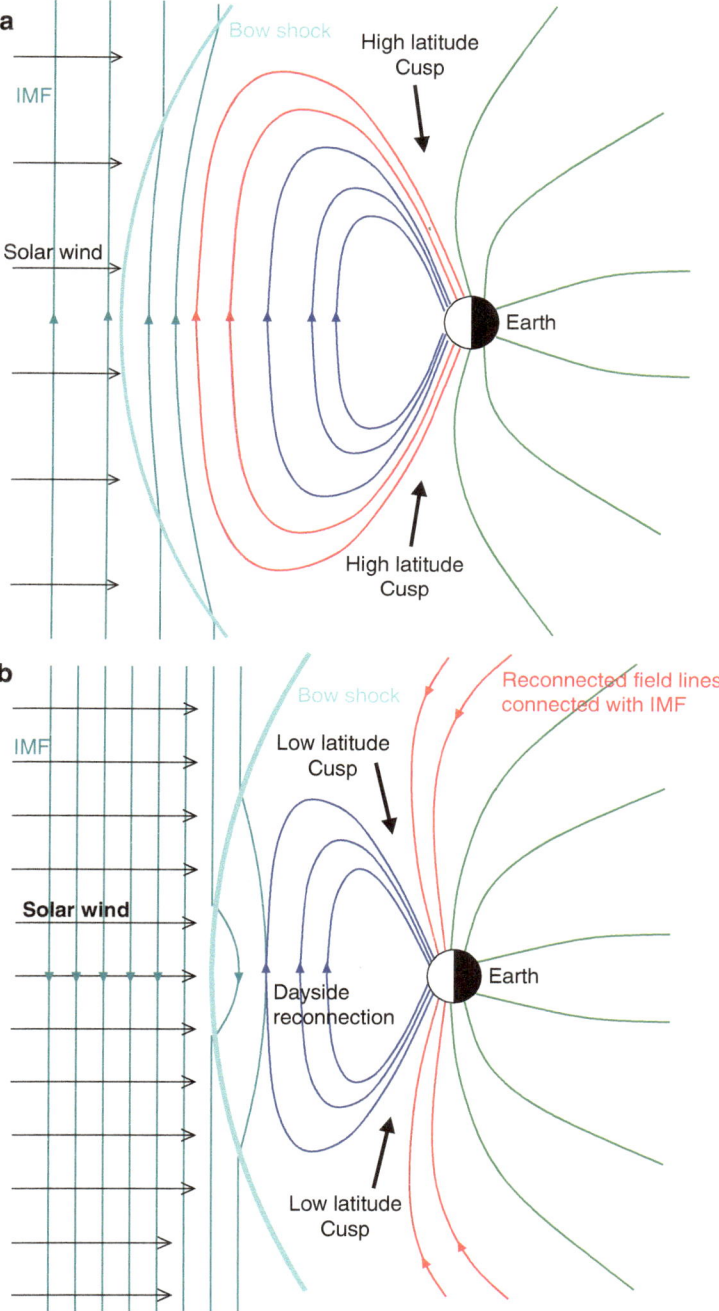

Fig. 1.3 (continued)

The combination of these two effects disturbs the geomagnetic field, or in other words they increase the geomagnetic activity causing a magnetic storm. The rate of magnetic reconnection between the CME and the geomagnetic field is by far the most significant driver of geomagnetic activity. Geomagnetic activity is measured in a number of ways, most commonly the Dst [28] and K_p indices.

1.8 Summary

A coronal mass ejection, or CME, is a large eruption of magnetic field and plasma from the Sun. CMEs may contain masses in excess of 10^{13} kg, achieve speeds greater than 1,000 km/s and may span several 10's of degrees heliographic latitude and/or longitude. They are most likely ejected in explosive form because of the need to release vast quantities of stored magnetic field complexity and energy from the solar environment. As they evolve through interplanetary space they accumulate solar wind material, and although their original structure and composition is likely retained, some distortion can arise from the interaction with the surrounding solar wind. Some components of the CME (e.g., the filament) may not survive the journey to 1 AU while others (e.g., a forward shock) may develop en-route. They are observed with imagers, which observe the CME and its related phenomena close to the Sun, and with in-situ spacecraft, which measure the properties of the CME as they pass through it. In the last decade it has been possible to directly track CMEs continually through the inner heliosphere by way of heliospheric imaging.

When they occasionally impact the Earth, CMEs alter the behavior of the Earth's magnetosphere. If the magnetic field of the arriving CME has a strong southward component, magnetic reconnection between it and the dayside geomagnetic field may occur, resulting in an opening of field lines and a large injection of particles into the magnetosphere. The increased pressure and/or shock from the CME may cause a compression of the magnetosphere as well, but this is insignificant compared to the disruption caused by reconnection. These result in large disturbances to the Earth's magnetic field known as a (geo)magnetic storm. Such storms are known to cause a variety of potentially serious deleterious effects.

Studying CMEs is therefore important not only from a scientific basis, but also for technical interests. Scientifically, CMEs provide information on the evolution of the Sun through a crucial process by which it removes built-up energy stored in the complexity of its magnetic field. Technically, CMEs are an obstacle to the

Fig. 1.3 (continued) Diagram representing a simplified version of the behavior of high-latitude geomagnetic field lines: (**a**) under low geomagnetic activity conditions, where the cusp region is at high latitudes; (**b**) when a southward-directed magnetic field arrives at the Sun, magnetic reconnection enables more dayside field lines to open, moving the cusp region towards the equator resulting in an aurora being observed at lower latitudes. Increased pressure also reduces the size of the magnetosphere, further enhancing this effect. See also Fig. 4.4

continuing development of electrical and space technology, so their understanding is crucial not only to assist in the design of "storm resistant" hardware, but also in the prediction of their arrival and the consequences to space weather at the Earth when they do.

References

1. Baker, D.N., Balstad, R., Bodeau, J.M., Cameron, E., Fennel, J.F., Fisher, G.M., Forbes, K.F., Kintner, P.M., Leffler, L.G., Lewis, W.S., Reagan, J.B., Small, A.A., III, Stansell, T.A., Strachan, L., Jr.: Severe Space Weather Events – Understanding Societal and Economic Impacts, NRC Workshop Rep. (2009)
2. Bame, S.J., Asbridge, J.R., Feldman, W.C., Fenimore, E.E., Gosling, J.T.: Solar Phys. **62**, 179–201 (1979)
3. Borrini, G., Gosling, J.T., Bame, S.J., Feldman, W.C.: J. Geophys. Res. **87**, 7370–7378 (1982)
4. Borrini, G., Gosling, J.T., Bame, S.J., Feldman, W.C.: Solar Phys. **83**, 367–378 (1983)
5. Cane, H.V., Kahler, S.W., Sheeley, N.R., Jr.: J. Geophys. Res. **91**, 13321–13329 (1986)
6. Carrington, R.C.: Mon. Not. R. Astron. Soc. **20**, 13–15 (1860)
7. Chen, P.F.: Living Rev. Solar Phys. **8**, 1 (2011)
8. Dasso, S., Mandrini, C.H., Démoulin, P., Luon, M.L.: Astron. Astrophy. **455**, 349–359 (2006)
9. DeForest, C.E., Howard, T.A., McComas, D.J.: Astrophys. J. **769**, 43 (2013)
10. Dryer, M.: Space Sci. Rev. **67**, 363–419 (1994)
11. Emslie, A.G., Kucharek, H., Dennis, B.R., Gopalswamy, N., Holman, G.D., Share, G.H., Vourlidas, A., Forbes, T.G., Gallagher, P.T., Mason, G.M., Metcalfe, T.R., Mewaldt, R.A., Murphy, R.J., Schwartz, R.A., Zurbuchen, T.H.: J. Geophys. Res. **109** (2004). doi:10.1029/2004JA010571
12. Fenimore, E.E.: Astrophys. J. **235**, 245–257 (1980)
13. Fetter, A.L., Walecka, J.D.: Theoretical Mechanics of Particles and Continua, Chap. 9. McGraw-Hill, New York (1980)
14. Forbes, T.G., Linker, J.A., Chen, J., Cid, C., K'ota, J., Lee, M.A., Mann, G., Mikić, Z., Potgieter, M.S., Schmidt, J.M., Siscoe, G.L., Vainio, R., Antiochus, S.K., Riley, P.: Space Sci. Rev. **123**, 251–302 (2006)
15. Forsyth, R.J., Bothmer, V., Cid, C., Crooker, N.U., Horbury, T.S., Kecskemety, K., Klecker, B., Linker, J.A., Odstrcil, D., Reiner, M.J., Richardson, I.G., Rodriguez-Pacheco, J., Schmidt, J.M., Wimmer-Schweingruber, R.F.: Space Sci. Rev. **123**, 383–416 (2006)
16. Gloeckler, G., Fisk, L.A., Hefti, S., Schwadron, N.A., Zurbuchen, T.H., Ipavich, F.M., Geiss, J., Bochsler, P., Wimmer-Schweingruber, R.F.: Geophys. Res. Lett. **26**, L157–L160 (1999)
17. Graham, G.: Philos. Trans. R. Soc. Lond. **383**, 96–107 (1724)
18. Hargreaves, J.K.: The Solar-Terrestrial Environment. Cambridge Atmospheric and Space Science Series, vol. 5. Cambridge University Press, Cambridge (1992)
19. Howard, T.: Coronal Mass Ejections, An Introduction. Springer, New York (2011)
20. Howard, T.A., DeForest, C.E.: Astrophys. J. **746**, 64–75 (2012)
21. Hundhausen, A.J.: J. Geophys. Res. **98**, 13177–13200 (1993)
22. Hundhausen, A.J., Burkepile, J.T., St. Cyr, O.C.: J. Geophys. Res. **99**, 6543–6552 (1994)
23. Illing, R.M.E., Hundhausen, A.J.: J. Geophys. Res. **90**, 275–282 (1985)
24. Lepri, S.T., Zurbuchen, T.H., Fisk, L.A., Richardson, I.G., Cane, H.V., Gloeckler, G.: J. Geophys. Res. **106**, 29231–29238 (2001)
25. MacQueen, R.M.: Philos. Trans. R. Soc. Lond. A. **297**, 605–620 (1980)
26. Odenwald, S.F.: The 23rd Cycle. Columbia University Press, New York (2001)
27. Roelof, E.C., Decker, R.B., Kimigis, S.M.: Proceedings of Solar Wind 12, 1216, pp. 359–362. Saint-Malo, France (2010)

28. Sugiura, M.: Ann. Int. Geophys. Year **35**, 945–948 (1964)
29. van Houten, C.J.: Bull. Astron. Soc. Neth. **11**, 160–163 (1950)
30. Webb, D.F., Cheng, C.-C., Dulk, G.A., Edberg, S.J., Martin, S.F., McKenna-Lawlor, S., McLean, D.J.: In: Sturrock, P.A. (ed.) Solar Flares: A Monograph from Skylab Workshop II, p. 471. Colorado Associated University Press, Boulder (1980)
31. Yashiro, S., Gopalswamy, N., Michalek, G., St. Cyr, O.C., Plunkett, S.P., Rich, N.B., Howard, R.A.: J. Geophys. Res. **109** (2004). doi:10.1029/2003JA010282

Chapter 2
History

In this chapter we briefly review the history of CME observation and the scientific contributions to our understanding of these phenomena from those observations. Less emphasis has been placed on what the author considers to be secondary effects of the CME (e.g., radio bursts, solar surface activity, solar energetic particles). This is not to say that the study of these secondary effects has not made significant contributions to our understanding of the CMEs and to solar physics in general, much of which predates the discovery of the CME. In this chapter we address them only in their early historical context, and mostly before the CME was directly observed since this was the way by which CMEs were eventually identified.

2.1 Early Years of Solar and CME Observation

Early studies of the solar corona date back to antiquity: the first recording of the solar eclipse was in 1223BC [229]; and the first identification of the solar corona was probably in 968AD [73]. The first observation of a CME, however, probably didn't occur until the mid-nineteenth century when the corona was being studied in great detail for the first time. Until the coronagraph was invented in the 1930s, coronal studies were limited only to during solar eclipses when the moon temporarily blocked out the bright solar photosphere. It is therefore not surprising that CMEs, which only occur a few times day, were not observed during previous eclipses, which only last a few minutes when observed from any fixed location. The first picture of a CME is probably from the drawing reproduced in Fig. 2.1. This is a drawing of the eclipse observed on 18 July 1860 in Torreblanca (Spain). Toward the southwest (lower-right) of the image appears to be a bubble-shaped structure that is disconnected from the Sun and remaining corona. Drawings of the same eclipse by other workers also reveal an extended structure in this region of the Sun. Such a structure in the solar corona would not be observed again for 100 years, and none realized the significance of that observation at the time.

T. Howard, *Space Weather and Coronal Mass Ejections*, SpringerBriefs in Astronomy,
DOI 10.1007/978-1-4614-7975-8_2, © Timothy Howard 2014

Fig. 2.1 Drawing of the 1860 eclipse recorded by Tempel [174] and identified later by Jack Eddy. This is believed to be the first observation of a coronal mass ejection

In 1852, Edward Sabine and other workers "absolutely" connected the 11-year sunspot cycle with geomagnetic activity [183,186]. In 1859 a powerful flare erupted from a large active region on the Sun and 18 h later the most intense recorded magnetic storm in history, the Carrington Event (Carrington Storm [22]) occurred at Earth, causing telegraph systems to fail across Europe and North America (see Sect. 1). Contemporary estimates of the Dst index for the Carrington Event range from −1,600 [205] to −850 nT [191]. Associations between flares and geomagnetic storms continued in the following decades, although the relationship was not one-to-one. For example, Maunder [151] and Greaves and Newton [58,59] showed that the most intense geomagnetic storms were usually accompanied on the Sun by groups of large-area sunspots. In 1931, Chapman and Ferraro [24–26] proposed that this correlation could be explained if there was a sporadic ejection of ionized material from the Sun.

2.2   Early Ground Measurements of CME Signatures

In the decades leading to the space age, investigations of the interplanetary counterparts of solar eruptions were being investigated via the energetic particles being accelerated by them and arriving at the Earth, and via their radio signatures. Scott Forbush in 1946 noted bursts of cosmic ray intensity at the Earth that he associated with energetic particles arriving from the Sun [40]. Solar energetic particles, or SEPs were immediately associated with solar flares and with variable magnetic fields around sunspots [41]. These ground level cosmic ray enhancements, or GLEs (ground level enhancements), were later detected by neutron monitors in 1956 [154] and riometers in 1959 [175]. SEPs were described by John Wild and co-workers in 1963 to be accelerated by two stages: Flare acceleration up to $\sim$100 keV of electrons, and then a second-level acceleration process caused by an outward-moving fast magnetohydrodynamic (MHD) shock [221]. This two-stage acceleration was confirmed using in-situ observations through the 1980s and 1990s [60].

The construction of the first radiospectrograph at Penrith in New South Wales (Australia) in 1950 enabled the study of solar-related radio bursts. These were classified into three types: Type I bursts were short-lived, narrowband, and occurred during magnetic storms; Type II bursts were longer in duration, accompanied solar flares at the Sun and drifted gradually in frequency; while Type III bursts were short-lived and broadband, where the frequency of maximum intensity drifted rapidly [217]. A fourth type of burst event, designated Type IV, was identified later in 1957 by workers using an interferometer at the Nançay observatory in France [7]. They were long-duration, associated with solar flares, and often followed a Type II burst. Type V bursts, which often followed Type III, were identified in 1959 [220]. From McLean and Labrum [153]:

> The observations of Type II and Type III bursts contributed significantly to the developing subject of solar flare 'anatomy' (Wild et al. 1954b) [219]. It was found repeatedly that groups of Type III bursts occurred at the very start of flares, coincident with the arrival of X-rays as signified by the onset of sudden ionospheric disturbances. The Type II burst, if one occurred, began some minutes later. (pp. 12–13 [153])

It is from Type II radio bursts that the first height-time plots of CMEs were indirectly plotted. Figure 2.2 provides an example of such a plot from Wild et al. [218]. Type II bursts are manifestations of interplanetary shocks, and so these measurements were related to the shock rather than the CME itself. Nonetheless, these measurements did provide indicators of CME speeds in the heliosphere 20 years before their discovery in coronagraphs.

2.2.1 CME Observations in the Early Space Age

The launch of *Sputnik-1* in 1957 began the space age but CME signatures in inter-planetary space were not measured until spacecraft left the Earth's magnetosphere. It was expected that interplanetary forward shocks would be observed, following a suggestion by Thomas Gold in 1955 that high-speed plasma ejected from the Sun would produce a collisionless shock in the interplanetary medium [46]. Indeed, a forward shock was directly observed in interplanetary space in 1962 by the *Mariner 2* spacecraft [192]. A further two were later reported in 1968 by Jack Gosling and

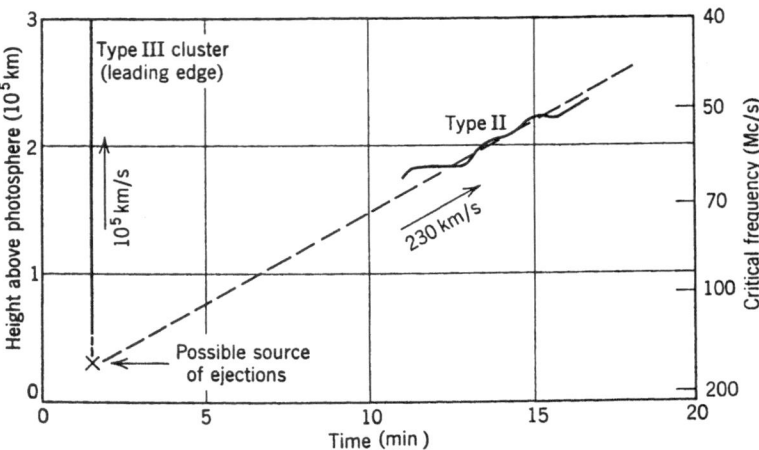

Fig. 2.2 An early example of a height-time plot, derived from Type II and Type III bursts [218]. This plot results in a speed of 230 km/s

co-workers using the *Vela 3* spacecraft pair [51]. Hundhausen et al. [106] used solar wind observations of shock disturbances to estimate that a large shock was associated with an ejection of 10^{13} kg and 10^{32} ergs from the Sun. By the year that the CME was discovered in coronagraphs (1973), several publications had emerged reporting interplanetary shock observations [34, 103, 104, 112, 113, 135, 167, 201], and many were associated with geomagnetic activity. Hence by the discovery of the CME, the theory of the formation and propagation of interplanetary shocks was firmly established, and had been confirmed with direct observation using in-situ spacecraft. They were associated with eruptions from the Sun (then mostly believed to be solar flares), and were known to cause increases in geomagnetic activity, particularly in the form of a sudden-(storm)-commencement, or S(S)C.

The first measurements of the ionic composition of the plasma following interplanetary shocks, made in 1972, revealed a helium abundance enhancement (HAE) [81]. The association of HAEs with solar flares had already been made a few years prior [4, 134]. Following a statistical study in which 73 cases of helium abundance enhancements (HAEs) were measured, it was suggested by Borrini et al. [8, 9] that HAEs were the interplanetary signatures of CMEs. High ionization states of oxygen and iron following interplanetary shocks were detected in 1979 [5, 38]. This provided information on the thermal state of CMEs, indicating that they were typically hotter than the surrounding solar wind. Other ions and temperature measurements followed, including magnesium and neon, further suggesting the presence of filament material or dense plasma from the low corona or chromosphere [9].

2.2.2   *Interplanetary Scintillation*

Sometime in the 1960s (probably 1964 [77]), Tony Hewish and co-workers at Cambridge University discovered that radio signals from distant sources (of the order of 100 MHz) fluctuated as a result of variations in the interplanetary medium. They showed that this interplanetary scintillation (IPS) can monitor the density of the solar wind (for a review of early work, refer to W. A. Coles [28]. Chapter 6 of the introductory text on CMEs by the author [93] provides a review of IPS.) By the time of discovery of the CME in 1973, several papers on this detection had appeared [76, 83–85]. It was not known at the time whether the transients observed were the same ones observed in the low corona, but it was clear that these were dense structures moving through the interplanetary medium between the Sun and the Earth.

By 1978 several interplanetary transients had been detected using IPS. Houminer and Hewish [84,85] investigated density enhancements in the interplanetary medium that were at low solar latitudes and appeared to co-rotate with the Sun. Watanabe and co-workers [210,211] reported on disturbances in the interplanetary medium which they attributed to flare shock waves. Interplanetary scintillation and proton density observed by the *Pioneer 6* and *7* spacecraft from January to April 1971 were found to be strongly correlated by Houminer and Hewish [86], and a relationship with the geomagnetic A_p index was also confirmed [86, 206–208]. Three transients were identified using IPS by Rickett [180] and these were correlated with *Pioneer 9* and *HEOS-2* at the Earth [180]. It should be noted that we use the word "transient" in a cautionary fashion here. Many of these early events were more likely the result of enhanced density regions of the Sun brought about by the merger of fast and slow solar wind streams, phenomena now known as corotating interaction regions (CIRs). Such structures corotate with the Sun but are time-stationary in structure, and so the term "transient" does not really apply. Vlasov identified two types of large-scale perturbations moving away from the Sun from 0.3 to 1.2 AU away from the Sun; those which vary over times of the order of 24 h (CMEs), and those that existed for several days (CIRs) [206].

Positive identifications of CMEs using IPS were made from 1978, when the radio telescope in the UK was upgraded. This enabled more radio sources to be monitored, thereby increasing the spatial resolution of the maps produced by IPS. This, along with coronagraph observations that were available by that time, allowed the first comparison of IPS transients with coronagraph CME images. Figure 2.3 shows an IPS map with a CME from 19 September 1980 from Tappin [195]. Each square on this figure represents a radio source and the red squares are those from which an increase in density has been identified. Later papers by Hewish and co-workers appeared that associated IPS CMEs with solar surface features, but most of those were low-latitude coronal holes [74, 75, 87, 198]. Interplanetary shock studies using IPS were studied through 1981–1985 by Woo and co-workers, who associated them with blast waves from solar flares [224–226].

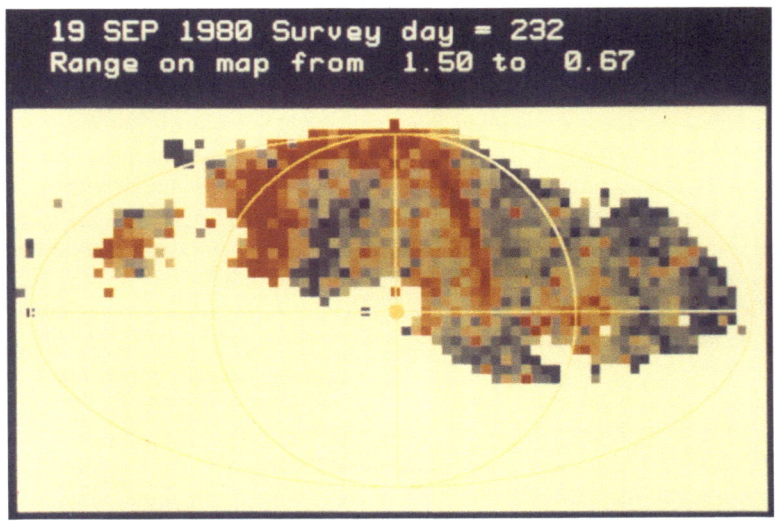

Fig. 2.3 IPS map of a CME from 19 September 1980. This is a Mollweide map with the Sun at the center and the 90° contour shown. Each *square* represents a radio source detected with the 3.6 Hectare Array and a *red square* indicates an increase in density (i.e., part of a possible CME) [195]

2.3   The Discovery of CMEs (1973)

The coronagraph was invented in 1931 by Bernard Lyot, allowing the continuous monitoring of the corona without the need to wait for a solar eclipse. This was achieved by permanently blocking the brighter light from the photosphere using a disk, known as an occulting disk [147]. It was not until decades later, however, that the sensitivity of the instrument was reduced to a level where faint coronal eruptions could be observed. This required spacecraft observation to identify them clearly, and workers then returned to ground data and identified them post-facto. The discovery of the CME is accredited to Richard Tousey, who identified transients moving with speeds of 400–1,000 km/s in the K-corona using *OSO-7* coronagraph observations. The discovery is reported in a review published in the Proceedings of the Fifteenth Plenary Meeting of COSPAR (*Space Research XIII*) in 1973 [203].

   At around the same time, coronal disturbances were being monitored using the ground coronagraph at Sacramento Peak in New Mexico. These were reported by Howard DeMastus, Bill Wagner and Rich Robinson in the *Solar Physics* journal in 1973. They refer to a number of "fast green line events" or "coronal transients" observed on the solar limb from 1956 to 1972, and they attempted to associate them with other forms of solar limb activity [32]. By this time coronal transients had also been recognised by workers using the Mauna Loa coronagraph, with observations published the following year [44, 132]. It seems likely that all groups had observed manifestations of the same phenomenon.

The first direct association between interplanetary shocks and CMEs was made by Gosling et al. [53] by comparing a CME observed by the coronagraph on *Skylab* with an interplanetary shock detected by *Pioneer 9*. Other early works include Dryer [33], Burlaga et al. [15] and Michels et al. [156].

A total of 20 CMEs were observed by *OSO-7* before it re-entered the Earth's atmosphere in 1974 [89]. The US space station *Skylab* was launched in 1973, which carried a coronagraph. Around 77 transients were observed by *Skylab* from May 1973 to February 1974 [162], and they were immediately identified as mass ejections [52]. The first appearance of the term "coronal mass ejection" appears to be in Gosling et al. [54], although the term "mass ejection coronal transient" appears in Hildner [80]. Initially, workers preferred to adhere to the more conservative "coronal transient", and the coronal mass ejection term was initially reserved for a particular type of eruption observed, but over time this term was used in more general contexts. By 1990, virtually all workers were referring to all large ejecta observed with a coronagraph as a coronal mass ejection or CME.

Observations of CMEs continued into the 1980s with the launch of the US Department of Defense Test Program satellite *P78-1* in February 1979, and of NASA's *Solar Maximum Mission* (*SMM*) in February 1980. On board each, amongst an assortment of other solar instruments, was the Naval Research Laboratory's coronagraph *Solwind* [155], and NASA's coronagraph/polarimeter C/P [148]. Among the discoveries of this next generation of space-based coronagraphs was the first Earth-directed CME by Russ Howard and co-workers in [90]. The term "halo CME" arises from this publication. The "classic" three-part CME structure shown in Fig. 1.1 was also first identified by the *SMM* C/P in this era [111]. Figure 2.4 shows images of CMEs obtained by these early instruments.

This second generation of spacecraft coronagraph collectively observed over 2,000 CMEs, thereby enabling the detailed statistical analysis of their properties for the first time. Such studies include those by Hundhausen et al. [107], who reported that the location of the CME was more evenly distributed around the Sun than those observed by *Skylab* (which were localized around the equator), and Howard et al. [91] and Hundhausen et al. [108], who surveyed almost a thousand CMEs over 3 years and provided extensive statistical results on structure, mass, angular span, location and kinetic energy. Hence, by 1995 solar physicists had a good picture of CME occurrence, structure, speed, mass and energy via a detailed investigation of single events as well as statistical surveys.

2.3.1   *Magnetic Clouds*

In Sect. 1.4, we discussed the internal core of the CME as comprising a flux rope that is often measured near 1 AU as a highly structured magnetic spiral. This signature was first identified in 1981, when Len Burlaga and co-workers identified a smoothly rotating magnetic field vector following an interplanetary shock for a single CME

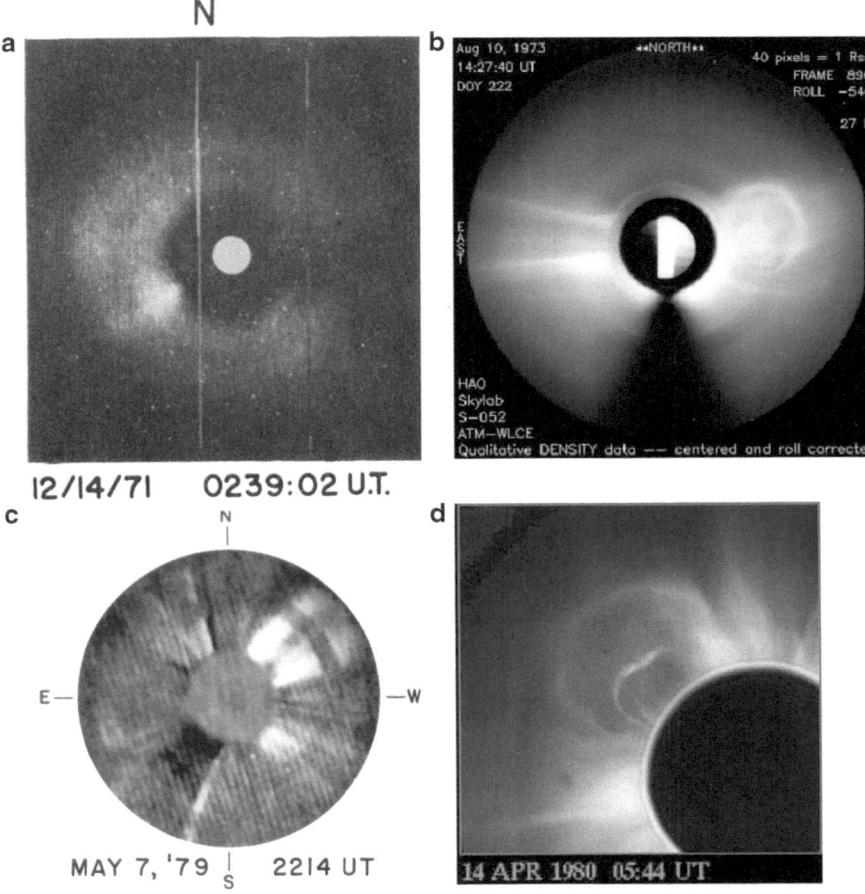

Fig. 2.4 Images of some of the early CMEs observed by space-based coronagraphs. (**a**) One of the first CMEs observed with *OSO-7* by Tousey [203]. This image was obtained on 14 December 1971 (Copyright Wiley-VCH Verlag GmbH & Co. KGaA. Reproduced with permission). (**b**) The coronagraph on board *Skylab* (available courtesy of the High Altitude Observatory (HAO)), obtained on 10 August 1973. Images from (**c**) *Solwind* on 7 May 1979 [121] (Reproduced with kind permission of Springer Science and Business Media), and (**d**) C/P on 14 April 1980 [193] (courtesy of HAO) follow

observed with five separate spacecraft (*Voyager 1* and *2*, *Helios 1* and *2* and *IMP-8*) [14]. They called it a "magnetic cloud" following a term used in theoretical studies starting from the 1950s [159]. Figure 2.5 shows their diagram of this event, including the structure that later became synonymous with CMEs observed in-situ: a shock, followed by a sheath, followed by the magnetic cloud (see Sect. 1.5). An accompanying paper [130], which presented a study of 45 magnetic clouds, made their connection with CMEs. These papers established the combination of

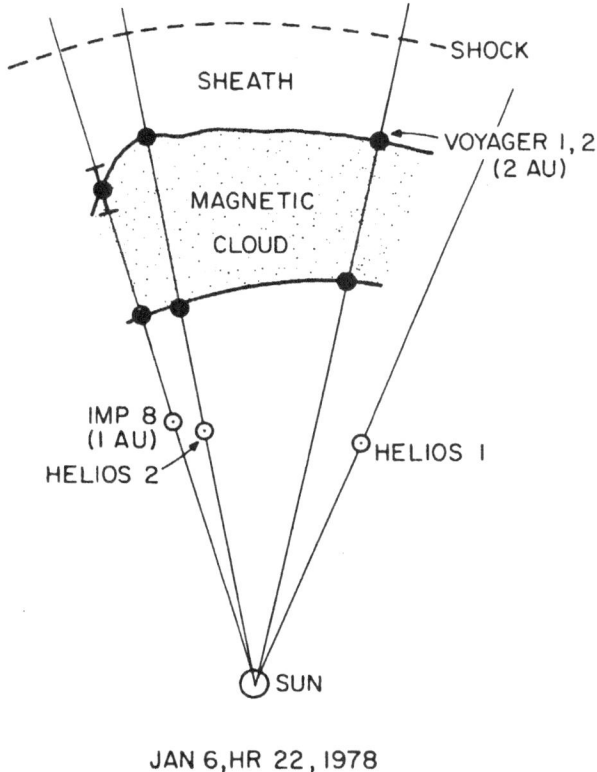

JAN 6,HR 22,1978
EQUATORIAL PLANE

Fig. 2.5 Sketch of the geometry of the magnetic cloud observed by Burlaga et al. [14] from spacecraft observations on 6 January 1978 (their Fig. 5). The *dots* show where the observed boundaries of the cloud would be at 22:00UT on that day, assuming they moved at constant speed

characteristics of magnetic clouds that are still used today: low temperatures, high magnetic field strength, and a smoothly rotating magnetic field vector. By 1990, it was accepted that magnetic clouds were probably manifestations of coronal mass ejections [13, 223], that they were regarded as a strong source of increased geomagnetic activity [227], and that they were often drivers of interplanetary shocks [130]. It was also known that only a subset of in-situ CMEs (30–50 %) showed a clear magnetic cloud signature [20, 48]. Other CMEs were identified by other signatures in the solar wind, such as the presence of an interplanetary shock, expansion signatures in the speed and density profiles, energetic particle and temperature decreases, and chemical composition signatures such as HAEs.

2.4   The Solar Flare Myth

In 1931, the same year that Chapman and Ferraro made their suggestion of the ionized material ejection as possibly being responsible for geomagnetic storms, Hale [62] suggested that this material came from large solar flares. Dellinger [35] associated flares with geomagnetic disturbances, while Newton [163, 164] found a statistical correlation between large flares and magnetic storms. Later in 1950, Chapman [23], who had not mentioned flares in his and Ferraro's initial suggestion of the cause of magnetic storms, then cited flares as the likely cause. Given the accumulation of evidence over the decades it is not unreasonable to conclude that solar flares were responsible for interplanetary and geomagnetic disturbances. Hence, when interplanetary shocks were first observed by spacecraft in the 1960s, they were assumed to be caused by solar flares, even though shock associations with flares were not always established [51, 103, 104, 192].

When the CME was finally discovered in the 1970s, it was also naturally assumed by many to be a shock wave from a solar flare. This assumption persisted despite early revelations that CMEs and geomagnetic storms were often not associated with flares [32, 52, 126, 148], and that the energy required to launch the mass ejection was much greater than that of the flare itself [148, 213].

While it was known that interplanetary shock waves were the likely cause of most geomagnetic sudden storm commencements, by the early 1970s some workers were already having doubts about their association with flares. Hundhausen expressed concerns in 1972 [103, 104], and workers using the early CME results from *Skylab* noted the inconsistency between CME and flare occurrence [52]. Joselyn and McIntosh [126] expressed surprise at the small percentage of flare-related geomagnetic storms, and Sime et al. [188] questioned the validity of describing a CME as a shock front with the observation that the flanks of the CME did not move laterally as the CME loop top moved outward. Further evidence was accumulated through the 1980s, including the movement of surrounding plasma ahead of the CME (implying that the CME cannot be a shock because the shock should be the leading feature) [187], the location of the flare as being at only one footpoint of the CME [66, 189, 190], and a lacking in coincidence in the timing between flare and CME onset [67–69]. Figure 2.6 shows two diagrams produced by Harrison in [64] demonstrating the relationship between the CME and its associated flare.

By 1992 the culmination of evidence from virtually every area of space physics research presented a strong case for a CME-centered, rather than flare-centered paradigm. In his excellent review in the *Annual Reviews of Astronomy and Astrophysics*, Steve Kahler [127] addressed the questions of

> [H]ow did we form such a fundamentally incorrect view of the effects of flares after so much observational and theoretical work... [and] what is the... evidence to support a primary role for CMEs? (p. 114 [127]),

by presenting evidence from CME and flare observations, metric radio bursts, interplanetary shocks and magnetic fields, solar-energetic particles and their geomagnetic consequences.

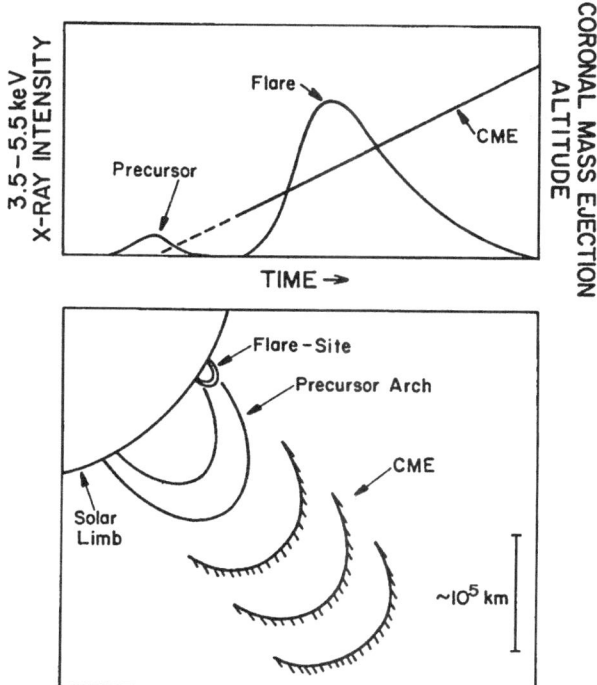

Fig. 2.6 Diagram representing the relationship between the CME and its associated flare (originally from Harrison [64] (his Fig. 6), and reproduced in Hundhausen [105] and Gosling [49]). Reproduced here with permission © ESO. The *top panel* shows the temporal relationship, showing the flare onset time occurring later than that of the CME, while the *bottom panel* shows the structural relationship, with the flare associated with one footpoint of the CME

Despite the solidity of evidence, most of the solar physics community continued to advocate the flare as the primary source of major space weather events. A review by Hudson in 1987 listed 42 great discoveries in solar physics and did not even mention CMEs [101], while a Lockheed Martin X-ray flare poster distributed at the AGU Fall Meeting in 1992 explicitly cited flares as the source for major geomagnetic storms. Finally, following a presentation of a soon-to-be traveling AGU exhibit addressing the Sun-Earth connection but not mentioning CMEs, Jack Gosling decided to write his now famous paper "The Solar Flare Myth", which was published in the *Journal of Geophysical Research* in late 1993 [49]. This paper confirmed that the source of interplanetary shocks and of most geomagnetic storms was the CME and not the flare, and that the relationship between the flare and CME was secondary (at best). He proposed a so called "modern paradigm" describing the relationship between flares, CMEs and geomagnetic activity.

The Solar Flare Myth paper caused outrage among the solar physics (particularly the flare) community and the debate intensified. A special session of the AGU

Meeting in Baltimore in May 1995 entitled 'Is "The Solar Flare Myth" Really a Myth?' was convened (a session to which Gosling himself was not invited). A challenge paper by Švestka [194] referred to Gosling's conclusions as "faulty and dangerous" and the response by Gosling and Hundhausen (p. 57) accused Švestka and others of attempting to re-classify the definition of a solar flare. A further response by Harrison [65] referred to the attempted reclassification to encompass virtually all eruptions from the Sun as "very misleading". Other challenges (e.g., Hudson et al. [102] and Pudovkin [172]) and responses were issued throughout 1995 and it seemed likely that this debate would remain unresolved for years to come.

Towards the end of 1995, however, the intensity of the Solar Flare Myth debate suddenly appeared to die away. Not coincidentally, the *SOHO* spacecraft was launched in December 1995 (Sect. 2.5.2). Perhaps the clarity of CME data from LASCO proved more conclusive, or perhaps the community was distracted by the quality of the data delivered by *SOHO*, but when the dust settled it appeared that the CME community had prevailed. The final settling of the debate probably occurred at a meeting on CMEs in Bozeman Montana in 1996 where a large number of those from the flare camp were present. Loren Acton, a main player in solar flares, was instrumental in getting the solar community to take notice. It is generally accepted today that CMEs, and consequently large transient solar wind disturbances and geomagnetic storms, are not caused by solar flares.

2.5   The 1990s

Only one spacecraft relevant to CME study (*SMM*) was launched in the 1980s, and had that been launched only 2 months earlier it would have been a 1970s launch. So while many of the spacecraft launched in the 1970s continued to function well into the 1980s (and some into the 1990s and later), after *SMM* no new missions of significance to CME study were launched throughout the 1980s decade. Furthermore, once *SMM* re-entered Earth's atmosphere in 1989, no continuous surveillance of the outer corona occurred until 1996. The next significant solar mission, *Ulysses*, was launched in 1990 although it was originally scheduled for launch in 1986.

2.5.1   *The Next Generation of In-Situ Probes:* Ulysses, Wind *and* ACE

Ulysses, launched in 1990, carried an assortment of magnetic field, energetic particle and other experiments, and was placed into a near polar orbit around the Sun. It achieved this by a gravitational assist around the planet Jupiter, and it needed to pass closer to the planet than any previous artificial object to do so. Given its unique orbit, the contributions to solar and interplanetary exploration made by *Ulysses* were mainly of observations of the polar regions of the Sun. Using *Ulysses*, Gosling [50] showed that CMEs can occur in the fast solar wind and several studies were made of CMEs at distances out to 5 AU [19,55,184,228]. Later, CMEs observed with *Ulysses* were compared with transients observed with IPS [124, 125] and with heliospheric white light images when they became available [196]. Charge state distributions of CMEs were investigated by Henke et al. [71, 72], who found that the charge state ratios of heavy solar wind ions (C^{6+}/C^{5+}, O^{7+}/O^{6+}, Si^{10+}/Si^{9+}, Fe^{12+}/Fe^{11+}) were related to the structure of the internal magnetic field.

In February 2008, *Ulysses* lost its secondary X-band transmitter which, among other things, enabled the regulation of temperature of the spacecraft. As it was on its way out away from the Sun, operators predicted 6 weeks before it froze to a point beyond operation. Despite this, the spacecraft continued to function for a further 18 months, and was finally turned off on 30 June 2009.

The next two solar/interplanetary in-situ spacecraft (*Wind*, launched November 1994 and *ACE*, launched August 1997) contained sophisticated instrumentation and have since been used as scientific and monitoring probes. The instruments on board *Wind* and *ACE* were mostly improved or modified versions of those already tested on previous missions. Likewise, the orbits from each spacecraft are not unlike those that had been seen before. The L1 location of *ACE* had been occupied by *ISEE-3* 20 years previously, and *Wind* was certainly not the first spacecraft to assume a high-Earth orbit passing into the upstream solar wind region beyond the magnetosphere.

Wind and *ACE* provided a continuous datastream of CME properties that remain continuous to this day. Their enhanced instrumentation also provided more in-depth studies of phenomena in the interplanetary medium. This, coupled with the next generation of imaging instruments (discussed in the following sections), allowed for the first time a reliable long-term continuous monitoring of solar, interplanetary and magnetospheric activity. This greatly enhanced our space weather forecasting. *ACE* remains the most crucial early-detection system for space weather, as it is always in the Sunward direction of the Earth. Unfortunately, it does not provide much of an early warning, as when a CME reaches *ACE* at L1 it is usually only around an hour away from arriving at Earth.

Statistical studies of CMEs using *Wind* and *ACE* include those by Cane and Richardson [21, 177], Lynch et al. [145, 146] and Howard and Tappin [94]. Iron charge distribution of CMEs were investigated by *ACE* by Gloeckler et al. [45] and Lepri et al. [141], who found typical charges of 9+ to 11+, but charges greater than 16+ were also identified . As with earlier studies, the higher charge states were attributed to hot plasma originating low in the solar corona or from initial heating during the launch of the CME. Later work includes the investigation of solar wind heating by CME-driven shocks and the relationship between composition and solar surface parameters [133, 176].

As mentioned previously, it was already known by 1990 that only a small fraction of CMEs were observed to contain magnetic clouds. However, by that year a global picture of the structure of a magnetic cloud had been formed ([17], see Fig. 2.7). Consequently, empirical modeling of magnetic clouds using in-situ data were based

Fig. 2.7 Sketch of a global
view of a magnetic cloud
through the ecliptic plane,
including the effects of solar
rotation [150]

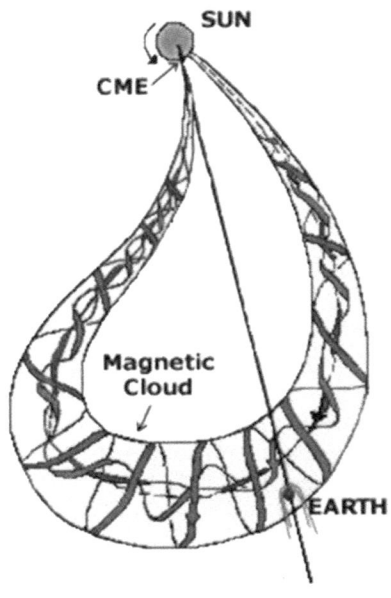

on this global picture. A model developed by Burlaga [12] and refined by Lepping et al. [139] using *ISEE-3* and *IMP-8* set the scene for magnetic cloud reconstruction techniques for when the next generation of in-situ data became available. Several different methods for such reconstruction followed; Riley et al. [181] provides a review of some of these models.

Since the launch of *Ulysses* in 1990, several hundred publications have appeared dealing with magnetic clouds with *Ulysses*, *Wind* and *ACE*. Studies have involved comparing them with CMEs [21, 177], solar surface structures [29, 137] and solar flares [173], geoeffectiveness [37, 61], magnetic reconnection [56, 57] and even internal whistler wave propagation [160]. The general picture of magnetic clouds remains as was defined from their discovery, but they are now an integral and essential part of CME study.

2.5.2 *The* SOHO *Era*

The *Solar and Heliospheric Observatory* (*SOHO*), launched in December 1995, was part of the European Space Agency's (ESA) "cornerstone" of its long-term "Horizon 2000" science program, and it was truly a cornerstone of solar physics research. First and foremost was the quality and variety of the data provided by its 12 instruments

Table 2.1 The 12 *SOHO* instruments in order of field of view. The instruments are identified first by their acronym then their full names, their field of view (if applicable), primary purpose and a reference to the instrument paper is also provided. Each of these papers were published in a special edition of the *Solar Physics* journal in 1995

Acron.	Name	Field of view	Primary purpose	Ref.
SWAN	Solar Wind Anisotropies	Whole-sky	Lyman alpha radiation detector	[6]
LASCO	Large Angle Spectroscopic Coronagraph	1.1–30 $R_\odot$	White light and EUV coronagraph	[10]
EIT	Extreme-Ultraviolet Imaging Telescope	Full solar disk	Multiwavelength EUV imager	[31]
MDI	Michelson Doppler Imager	Full solar disk	Solar oscillations and magnetic field investigation	[185]
UVCS	UltraViolet Coronagraph Spectrometer	$\sim$40 $\times$ 60′	UV spectroscopy and visible polarimetry studies	[131]
CDS	Coronal Diagnostic Spectrometer	$\sim$240 $\times$ 240″	EUV imaging spectrometer	[70]
SUMER	Solar Ultraviolet Measurements of Emitted Radiation	Thin slits	EUV analysis	[222]
CELIAS	Charge, Element, and Isotope Analysis System	N/A	Solar wind and particle detector	[88]
COSTEP	Comprehensive Suprathermal and Energetic Particle Analyzer	N/A	Energetic particle detector	[161]
ERNE	Energetic and Relativistic Nuclei and Electron Experiment	N/A	Energetic particle detector	[202]
GOLF	Global Oscillations at Low Frequencies	N/A	Helioseismology observer	[43]
VIRGO	Variability of the Solar Irradiance and Gravity Oscillations	N/A	Helioseismology and radiometry	[42]

(summarised in Table 2.1). While many of these types of instruments had been used in the past, on *SOHO* they were of higher quality, and were all available on board a single spacecraft.

With regard to CME study, the EUV imager EIT and spectrometer CDS provided invaluable information on solar eruptions associated with CMEs, but the major contributors to CME research were of course the coronagraphs. LASCO originally consisted of three coronagraphs, C1 with a field of view (FOV) of 1.1–3.0 $R_\odot$, C2

(FOV 1.5–6.0 $R_\odot$) and C3 (FOV 3.7–30 $R_\odot$). C2 and C3 are white light imagers, while C1 observed at variable EUV wavelengths.

For $2\frac{1}{2}$ years *SOHO* returned images of unprecedented detail on the Sun, including CMEs. The sensitivity of LASCO led to halo (Earth directed) CMEs being easily detected for the first time, and a large statistical database of CME observations had begun. Then on 25 June 1998, the spacecraft suddenly went into an uncontrollable spin and was lost for around a month. It was located on 23 July by a radio telescope and was dead in space, but careful analysis of its spin and trajectory enabled a prediction for when solar panels would be pointing at the Sun, providing power to the spacecraft. The first signal was received on 3 August and it was fully recovered by 16 September. Some ingenious engineering and scientific analysis went into the recovery of *SOHO*, including a study of the images of the Sun as they moved in and out of the field of view. The incident was later attributed to a sequence of operational errors leading to both gyroscopes being left off [204]. This is the only critical malfunction to occur in *SOHO* during its 17 year lifetime, and only the LASCO C1 camera was lost during the anomaly. Since January 2003 its data transmission capabilities have been limited following a malfunction in the pointing system of its high-gain antenna, which is now unable to move. Since the launch of *SDO* in February 2010, some of *SOHO*'s instruments have been turned off as they were deemed redundant (Sect. 5.1). No suitable replacement for LASCO has yet to be launched, and so its C2 and C3 coronagraphs remain operational and are central to the space weather forecasting effort.

The actual number of publications using LASCO is virtually impossible to identify, but it easily numbers in the thousands and probably tens of thousands. Along with the detailed study of CMEs, LASCO has assisted in research from solar wind origination to space weather to comet discovery. Two popular CME catalogs have appeared, managed by NRL (http://lasco-www.nrl.navy.mil/cmelist. html) and Goddard (http://cdaw.gsfc.nasa.gov/CME_list/). From 1996 to the end of 2011, the latter provided details on just under 18,000 CMEs observed with LASCO C2 and C3.

2.6 Heliospheric Imaging

Ten years ago, a new class of instrument was launched and a new era of CME observation was born. The heliospheric imager bridges the gap between coronagraph measurements near the Sun and in-situ measurements near the Earth by continually observing CMEs throughout their transit through the inner heliosphere. They function similarly to a coronagraph, in that they observe white light that has been Thomson scattered off free electrons in the heliosphere. The first of these, the Solar Mass Ejection Imager (SMEI), was launched in 2003 and was capable of observing the entire sky beyond around 20° from the Sun. A second instrument, simply called the Heliospheric Imager (HI) was launched on board *STEREO* in 2006.

2.6.1   *The* Helios *Zodiacal Light Cameras*

The heliospheric imager concept was proven by an instrument carried by the *Helios* spacecraft, launched much earlier in the mid-1970s. There were two spacecraft launched as part of this mission, *Helios 1* was launched in December 1974 and *Helios 2* was launched in January 1976. The missions ended in 1982 and 1976 respectively, although both spacecraft continued to deliver data until the mid-1980s. To this day they remain in their highly eccentric orbit about the Sun (perihelion ~0.3 AU, aphelion ~1.0 AU).

Each *Helios* spacecraft contained a white light imager as their zodiacal light experiment [138], which consisted of three photometers (white light cameras) oriented such that large strips at constant ecliptic latitude could be scanned as the spacecraft spun. The cameras were centered at 15°, 30° and 90° below the spacecraft equatorial plane, and the first two cameras scanned at 5.6°–22.5° longitude width, depending on the required angular resolution (see, for example, [121]). While it was not the primary science objective of the zodiacal light instrument, they did confirm that transients in the inner heliosphere could be detected in white light. Richter et al. [179] noted high-latitude "plasma clouds" and measured speeds for a number of them of around 300 km/s. They even associated one event with a CME observed by *Solwind* on 5 June 1979. This CME had a measured speed of 500 km/s and the *Helios* plasma cloud, observed on 6 June 1979, had a speed of 260–330 km/s, and therefore had decelerated en-route. Jackson and co-workers attempted to produce low-resolution images of these density changes [114,116,119,121]. The technique was developed further through the 1990s [79,117] and eventually *Helios* white light data were compared with IPS observations [78]. Other contemporary work compared *Helios* transients with coronagraph CMEs and interplanetary shocks [115,120,212,214], and by the end of the *Helios* era the association between coronagraph CMEs and white light and IPS CMEs had been firmly established.

2.6.2   *SMEI*

The Solar Mass Ejection Imager, SMEI [36, 122], was launched on the Air Force Space Test Program *Coriolis* spacecraft. SMEI consisted of three scanning cameras that built up an image of the sky throughout its 102 min polar orbit about the Earth. It observed the sky starting from around 20° elongation, or around 0.35 AU (75 $R_\odot$). While parts of the images were often contaminated by aurora and particle noise from the passage of the spacecraft through various regions of the magnetosphere (i.e., the polar caps and South Atlantic Anomaly), SMEI allowed for the first time direct measurement of complete CMEs at large distances from the Sun in white light. As with coronagraphs, the images were heavily projected, but unlike coronagraphs the problems posed by projection were mitigated with the application of geometry [96, 98, 197]. Hence, SMEI enabled the three-dimensional reconstruction of CMEs at large distances from the Sun. Figure 2.8 shows a SMEI image, shown as an all-sky Hammer-Aitoff projection.

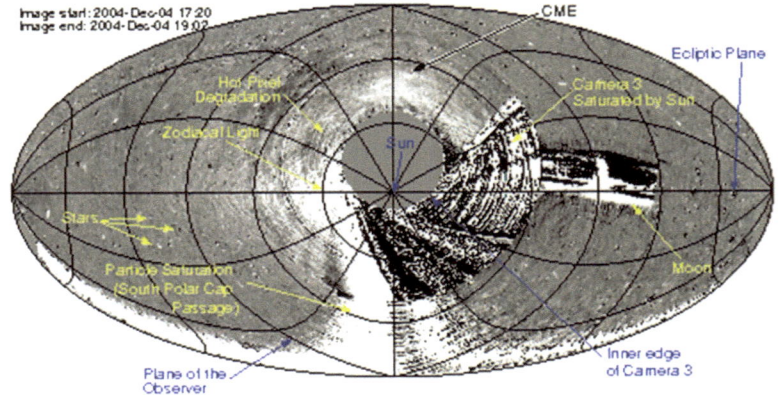

Fig. 2.8 A Hammer-Aitoff projection of a whole-sky SMEI image (it is probably more appropriate to think of it as a map) of a CME observed on 4 December 2004 around 18:00UT [200]. A number of features are labeled. Those labeled in *blue* indicate regions due to the projection, and those labeled in *yellow* are sources of interference. A CME appears in this map heading to the north (upwards), which is labeled

Early work with SMEI involved mostly height-time comparisons with coronagraph CMEs, interplanetary shocks and geomagnetic storms [99, 196, 199, 215] but some three-dimensional work was attempted from the start. For example, Jackson and co-workers extended their *Helios* and IPS reconstruction work to include SMEI data [123] and the author and co-workers have performed more simplified reconstruction techniques based on leading edge measurements of SMEI CMEs [99, 100]. Later, we developed more sophisticated 3-D reconstructions using the theory of Thomson scattering and the geometry of the CME [96, 98, 197]. Figure 2.9 shows some of these 3-D reconstructions of CMEs made using SMEI data.

SMEI was de-activated in September 2011, but work with its dataset continues.

2.7   *STEREO*

In 2006 the *STEREO* spacecraft [129] assumed an orbit and carried a suite of instruments never before seen on a solar mission. The purpose of *STEREO* was to provide multiple in-situ measurements and images of the Sun from different viewpoints from the traditional Sun-Earth line, and so each were placed in an orbit similar to that of the Earth about the Sun. The difference was that one spacecraft would orbit slightly faster than the Earth with the other slightly slower, resulting in leading and lagging spacecraft in the ecliptic plane. Figure 2.10 shows the location of each spacecraft at various times during the mission. The angular separation between the spacecraft and the Sun-Earth line grows by around 22.5° per year.

Along with providing continuous in-situ measurements of interplanetary transients, allowing a study of the longitudinal structure of CMEs [97, 100, 158], the *STEREO* imagers allow both the continuous monitoring of CMEs throughout their entire journey to 1 AU (including within 20° elongation that was missed by SMEI), and also a three-dimensional image of solar structures much in the same way as depth is perceived using our two eyes. Figure 5.4 shows combined views of the *STEREO* imaging suite, which is called SECCHI [92]. As the spacecraft separated further, three-dimensional reconstructions became possible with different instruments, first with the EUV low corona structures [3, 142], then the coronagraphs [95, 157, 158], and finally the heliospheric imagers [97].

The *STEREO* spacecraft continue to move apart and are functioning to date, even though they are now well on the far side of the Sun relative to the Earth. Eventually they will pass each other on the far side of the Sun and return from the opposite direction.

2.8   The Continuing Role of Past Missions

It is helpful to remind the reader that many of the missions launched in earlier years continue to function and play a role in our understanding of space and CMEs to this day. The *Voyagers*, for example, continue to monitor the outer regions of the heliosphere [30]. By the year 2000, they were over 58 AU from the Sun and were still capable of observing CMEs even there. At those distances, CMEs tend to merge with other dense regions (such as corotating interaction regions or other CMEs), the combination of which are called merged interaction regions (MIRs [16]). Richardson et al. [178] studied a single event from the Sun to the *Wind* spacecraft at 1 AU to *Ulysses* at 5 AU then to *Voyager 2* at 58 AU. Similar studies include the "Bastille Day" CME by Burlaga et al. [18] and a series of events the following year [209]. A review of Voyager observations of MIRs involving CIRs can be found in Lazarus et al. [136].

Along with the *Voyager* observations, new publications continue to emerge from ongoing missions dating from the 1970s, such as *IMP-8* [47, 140]. Also, analyses of data from spacecraft no longer operating continue to yield new scientific results, such as those from *Helios* [47, 118, 144], *ISEE-3* [39, 182], *Solwind* [27, 128] and *SMM* [11, 149].

2.9   Other Missions

We have focused in this chapter on those missions that were used to observe CMEs directly, but other missions have been launched in the meantime that provided insight into the onset and evolution of CMEs. The instruments on board these missions did not observe CMEs directly, but did observe CME-related phenomena,

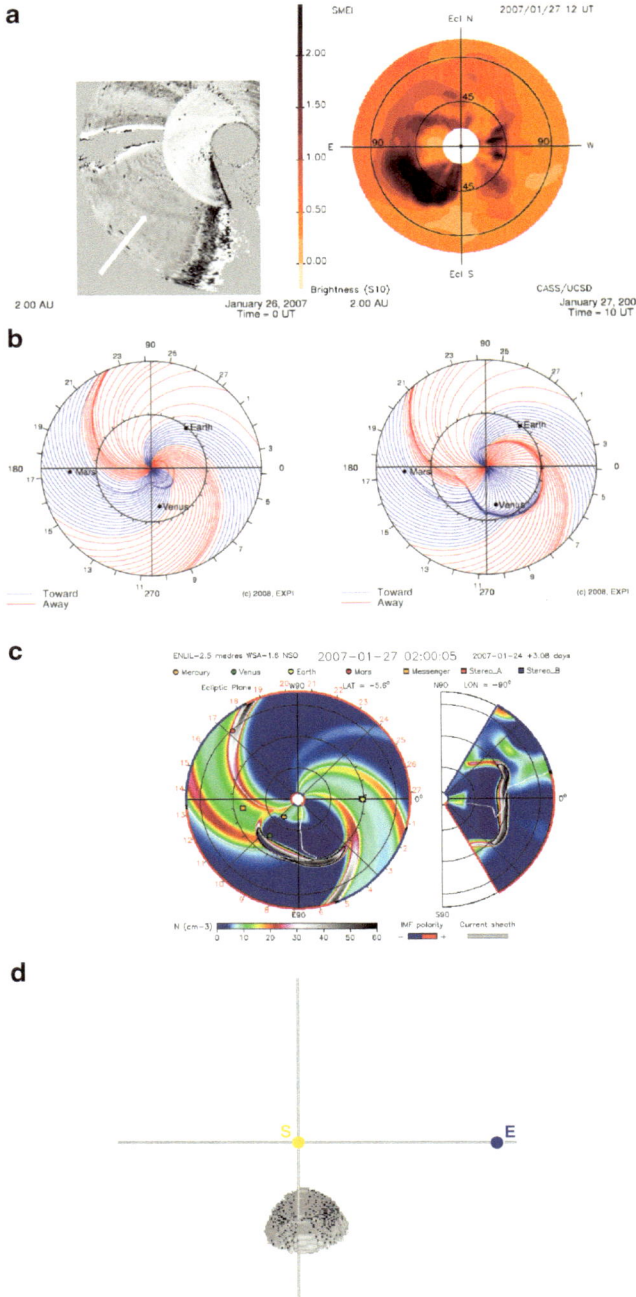

Fig. 2.9 Results from three-dimensional reconstruction efforts of CMEs that were used throughout the SMEI mission. These are for the CME observed in late January 2007 and panels (**a**)–(**c**) are from Webb et al. [216]. (**a**) SMEI "fisheye" image (*left*) and fisheye tomographic reconstruction

such as flares, filaments, and the response of the Earth's magnetosphere to their impact. More details on these missions are provided in Chaps. 2 and 3 of the author's introductory text on CMEs [93], but a list of many of these missions is provided in Table 2.2. We do not include the *GOES* spacecraft in this table, but a complete listing can be found in the Wikipedia entry for *GOES* (http://en.wikipedia.org/wiki/List_of_GOES_satellites). NOAA provides a website of the status of *GOES* (http://www.oso.noaa.gov/goestatus/) and a summary of the Space Environment Monitor (SEM) can be found at http://ngdc.noaa.gov/stp/satellite/goes/index/html.

2.10 Summary

To summarize, CMEs have been detected using a large variety of instruments and techniques. Directly:

1. Using white light coronagraphs that detect the light that is Thomson scattered from the electrons in the CME,
2. Directly measuring properties of the CME as it passes by in-situ spacecraft,
3. Measuring the changes in longwave radio signals from distant sources as the CME passes between them and the Earth (IPS);

and indirectly, through investigation of the secondary effects of CME launch and propagation:

1. Solar flares, observed in visible light, EUV, X-ray,
2. Erupting prominences/disappearing filaments, observed in visible light and EUV,
3. Other solar surface eruptions, such as post-eruptive arcades and coronal dimming,
4. Solar energetic particles accelerated by the shock in the interplanetary medium from the CME,
5. Type II and Type IV radio bursts, driven by the CME shock.

Figure 2.11 shows a timeline of the significant events that have led to an enhancement of our understanding of CMEs. The passage from ground-based to space-based observations is indicated, but the importance of the work leading up to the space age cannot be overstated. It seems clear that even by the time of the

----

Fig. 2.9 (continued) (*right*) at 12:00UT on 27 January. (**b**) HAFv2 (Sect. 3.2.2.2) solar wind disturbance reconstruction at 0UT on 26 Jan (*left*) and 10:00UT on 27 Jan (*right*), looking down onto the ecliptic plane and showing the orientation and polarity of interplanetary magnetic field lines. (**c**) ENLIL (Sect. 3.2.2.3) MHD solar wind injection reconstruction on 02:00UT on 27 Jan, showing the density of the solar wind and looking down onto the ecliptic plane (*left*) and along the north-south meridian (*right*). (**d**) TH model reconstruction on 27 Jan 2007, looking down onto the ecliptic plane. In panels (**b**)–(**d**) the location of the Sun and Earth are shown, and all four models show considerable agreement

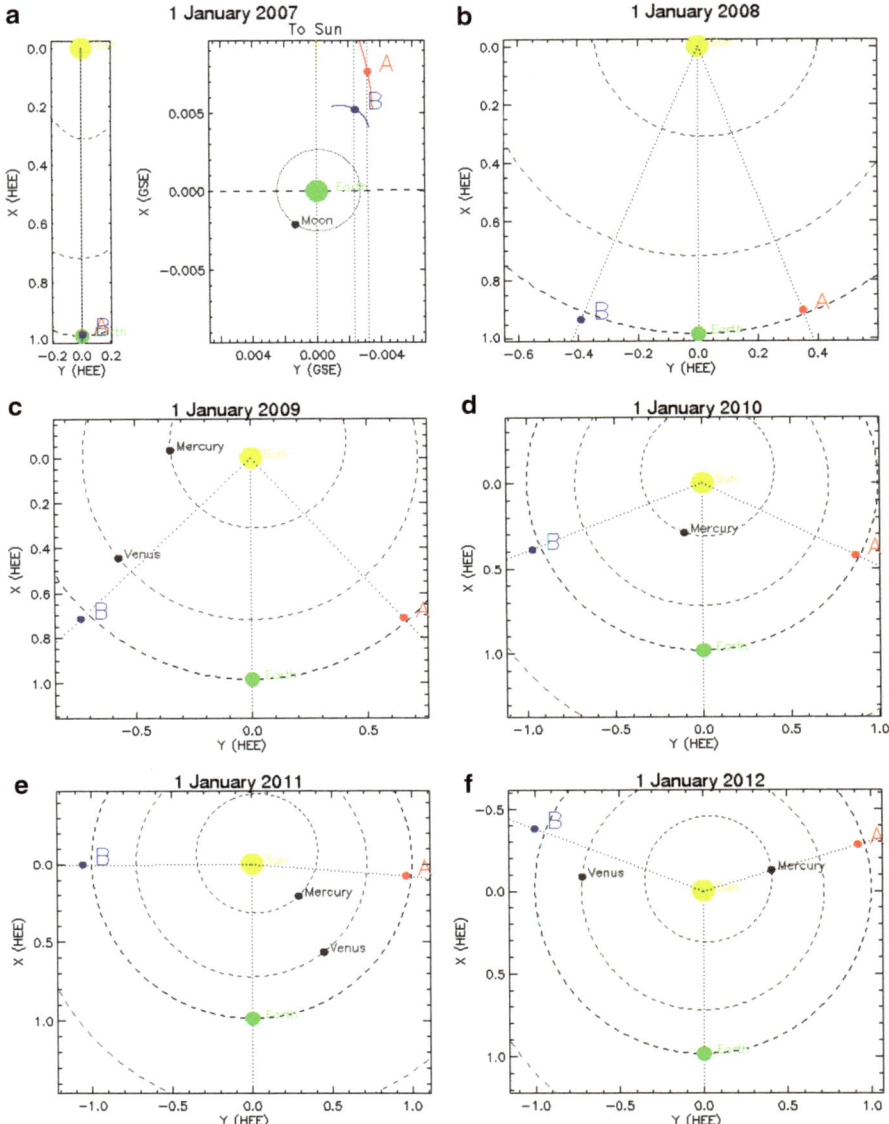

Fig. 2.10 The location of the *STEREO* spacecraft on 1 January of (**a**) 2007, (**b**) 2008, (**c**) 2009, (**d**) 2010, (**e**) 2011, (**f**) 2012 (from the STEREO website). The coloured circles indicate the following: *Yellow* = Sun, *green* = Earth, *red* = *STEREO-A*, *blue* = *STEREO-B*. The angular separation between the spacecraft and the Sun-Earth line grows by around 22.5° per day (Images provided courtesy of the "Where is STEREO?" tool (NASA/GSFC))

Table 2.2 Other missions that have not detected CMEs directly, but provided observations of CME-related phenomena, thereby providing insight into the nature of CMEs.

Name	Launch date	End date	Instrument	Ref.
IMP-8	26 October 1973	October 2001	In-situ suite	[170]
Voyager 1	5 September 1977	Continues to operate	Imagers, in-situ suite	[2]
Voyager 2	20 August 1977	Continues to operate	Imagers, in-situ suite	[2]
ISEE-3	12 August 1978	Became *ICE* in 1982	X-ray and gamma-ray spectrometers, in-situ suite	[168, 169]
Yohkoh	31 August 1991	14 December 2001	X-ray imagers, gamma-ray spectrometers	[152, 165, 166]
TRACE	1 April 1998	21 June 2010	UV and EUV imagers	[63, 109]
RHESSI	5 February 2002	Continues to operate	X-ray imager, spectrometer	[82, 143]
Hinode	22 September 2006	Continues to operate	Imaging suite	[1, 110]
SDO	11 February 2011	Continues to operate	Imaging suite	[171]

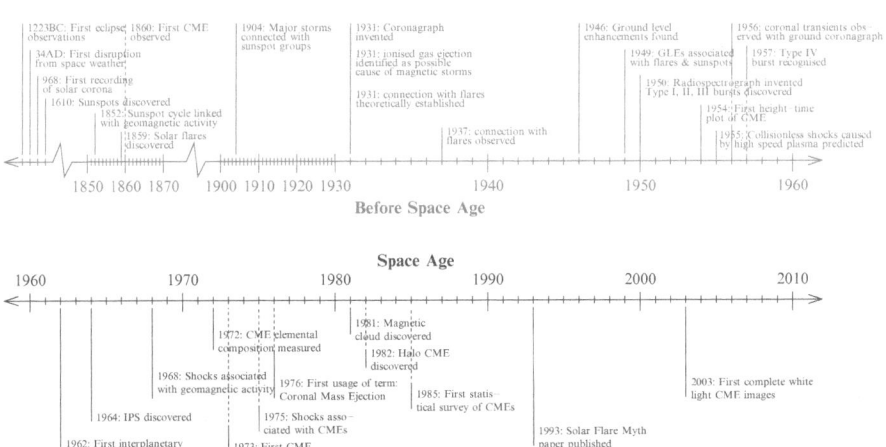

TIMELINE OF THE HISTORY OF CMES

Fig. 2.11 Timeline of the significant events that have led to an enhancement of our understanding of CMEs. It has been divided into before and during the space age (*green* and *blue* respectively)

emergence of the first spacecraft around 1960, our understanding of interplanetary transients and the interplanetary medium had a firm foundation.

References

1. Adams, M., Davis, J.M.: The Hinode (Solar-B) Webpage, Available via NASA/MSFC. http://solarb.msfc.nasa.gov/ (2009). Cited 10 Apr 2009
2. Angrum, A., Medina, E., Sedlacko, D.: The Voyager Webpage, Available via NASA/JPL. http://voyager.jpl.nasa.gov/ (2009). Cited 21 Jan 2009
3. Aschwanden, M.J., Wülser, J.-P., Nitta, N.V., Lemen, J.R.: Astrophys. J. **679**, 827–842 (2008)
4. Bame, S.J., Asbridge, J.R., Hundhausen, A.J., Strong, I.B.: J. Geophys. Res. **73**, 5761–5767 (1968)
5. Bame, S.J., Asbridge, J.R., Feldman, W.C., Fenimore, E.E., Gosling, J.T.: Solar Phys. **62**, 179–201 (1979)
6. Bertaux, J.L., Kyrölä, E., Quémerais, E., Pellinen, R., Lallement, R., Schmidt, W., Berthé, M., Dimarellis, E., Goutail, J.P., Taulemesse, C., Bernard, C., Leppelmeier, G., Summanen, T., Hannula, H., Huomo, H., Kehlä, V., Korpela, S., Leppälä, K., Strömmer, E., Torsti, J., Viherkanto, K., Hochedez, J.F., Chretiennot, G., Peyroux, R., Holzer, T.: Solar Phys. **162**, 403–439 (1995)
7. Boischot, A.: C. R. Acad. Sci. **244**, 1326 (1957)
8. Borrini, G., Gosling, J.T., Bame, S.J., Feldman, W.C.: J. Geophys. Res. **87**, 7370–7378 (1982)
9. Borrini, G., Gosling, J.T., Bame, S.J., Feldman, W.C.: Solar Phys. **83**, 367–378 (1983)
10. Brueckner, G.E., Howard, R.A., Koomen, M.J., Korendyke, C.M., Michels, D.J., Moses, J.D., Socker, D.G., Dere, K.P., Lamy, P.L., Llebaria, A., Bout, M.V., Schwenn, R., Simnett, G.M., Bedford, D.K., Eyles, C.J.: Solar Phys. **162**, 357–402 (1995)
11. Burkepile, J.T., Hundhausen, A.J., Stanger, A.L., St. Cyr, O.C., Seiden, J.A.: J. Geophys. Res. **109** (2004). doi:10.1029/2003JA010149 (2004)
12. Burlaga, L.F.: J. Geophys. Res. **93**, 7217–7224 (1988)
13. Burlaga, L.F., Behannon, K.W.: Solar Phys. **81**, 181–192 (1982)
14. Burlaga, L., Sittler, E., Mariani, F., Schwenn, R.: J. Geophys. Res. **86**, 6673–6684 (1981)
15. Burlaga, L.F., Klein, L., Sheeley, N.R., Jr., Michels, D.J., Howard, R.A., Koomen, M.J., Schwenn, R., Rosenbauer, H.: Geophys. Res. Lett. **9**, L1317–L1320 (1982)
16. Burlaga, L.F., McDonald, F.B., Goldstein, M.L., Lazarus, A.J.: J. Geophys. Res. **90**, 12127–12039 (1985)
17. Burlaga, L.F., Lepping, R.P., Jones, J.A.: In: Russell, C.T., Priest, E.R., Lee, L.C. (eds.) Geophysical Monograph Series, vol. 58, p. 373. AGU, Washington, D.C. (1990)
18. Burlaga, L.F., Ness, N.F., Richardson, J.D., Lepping, R.P.: Solar Phys. **204**, 399–411 (2001)
19. Burton, M.E., Smith, E.J., Goldstein, B.E., Balogh, A., Forsyth, R.J., Bame, S.J.: Geophys. Res. Lett. **19**, L1287–L1289 (1992)
20. Cane, H.V., Richardson, I.G., St. Cyr, O.C.: J. Geophys. Res. **102**, 7075–7086 (1997)
21. Cane, H.V., Richardson, I.G.: J. Geophys. Res. **108** (2003). doi:10.1029/2002JA009817
22. Carrington, R.C.: Mon. Not. R. Astron. Soc. **20**, 13–15 (1860)
23. Chapman, S.: J. Geophys. Res. **55**, 361 (1950)
24. Chapman, S., Ferraro, V.C.A.: Terr. Magn. Atmos. Electr. **36**, 77–97 (1931)
25. Chapman, S., Ferraro, V.C.A.: Terr. Magn. Atmos. Electr. **36**, 171–186 (1931)
26. Chapman, S., Ferraro, V.C.A.: Terr. Magn. Atmos. Electr. **37**, 147–156 (1932)
27. Cliver, E.W., Ling, A.G.: Astrophys. J. **556**, 432–437 (2001)
28. Coles, W.A.: Space Sci. Rev. **21**, 411–425 (1978)
29. Crooker, N.U.: J. Atmos. Solar Terr. Phys. **62**, 1071–1085 (2000)

30. Decker, R.B., Krimigis, S.M., Roelof, E.C., Hill, M.E., Armstrong, T.P., Gloeckler, G., Hamilton, D.C., Lanzerotti, L.J.: Science **309**, 2020–2024 (2005)

31. Delaboudiniére, J.-P., Artzner, G.E., Brunaud, J., Gabriel, A.H., Hochedez, J.F., Millier, F., Song, X.Y., Au, B., Dere, K.P., Howard, R.A., Kreplin, R., Michels, D.J., Moses, J.D., Defise, J.M., Jamar, C., Rochus, P., Chauvineau, J.P., Marioge, J.P., Catura, R.C., Lemen, J.R., Shing, L., Stern, R.A., Gurman, J.B., Neupert, W.M., Maucherat, A., Clette, F., Cugnon, P., van Dessel, E.L.: Solar Phys. **162**, 291–312 (1995)

32. Demastus, H.L., Wagner, W.J., Robinson, R.D.: Solar Phys. **100**, 449–459 (1973)

33. Dryer, M.: Space Sci. Rev. **33**, 233–275 (1982)

34. Dryer, M., Smith, Z.K., Endrud, G.H., Wolfe, J.H.: Cosm. Electrodyn. **3**, 184–207 (1972)

35. Delinger, J.H.: Terr. Magn. Atmos. Electr. **42**, 49–53 (1937)

36. Eyles, C.J., Simnett, G.M., Cooke, M.P., Jackson, B.V., Buffington, A., Hick, P.P., Waltham, N.R., King, J.M., Anderson, P.A., Holladay, P.E.: Solar Phys. **217**, 319–347 (2003)

37. Farrugia, C.J., Scudder, J.D., Freeman, M.P., Janoo, L., Lu, G., Quinn, J.M., Arnoldy, R.L., Torbert, R.B., Burlaga, L.F., Ogilvie, K.W., Lepping, R.P., Lazarus, A.J., Steinberg, J.T., Gratton, F.T., Rostoker, G.: J. Geophys. Res. **103**, 17261–17278 (1998)

38. Fenimore, E.E.: Astrophys. J. **235**, 245–257 (1980)

39. Feroci, M., Hurley, K., Duncan, R.C., Thompson, C.: Astrophys. J. **549**, 1021–1038 (2001)

40. Forbush, S.E.: Phys. Rev. **70**, 771–772 (1946)

41. Forbush, S.E., Gill, P.S., Vallarta, M.S.: Rev. Mod. Phys. **21**, 44–48 (1949)

42. Fröhlich, C., Romero, J., Roth, H., Wehrli, C, Andersen, B.N., Appourchaux, T., Domingo, V., Telljohann, U., Berthomieu, G., Delache, P., Provost, J., Toutain, T., Crommelynck, D.A., Chevalier, A., Fichot, A., Däppen, W., Gough, D., Hoeksema, T., Jiménez, A., Gómez, M.F., Herreros, J.M., Cortés, T.R., Jones, A.R., Pap, J.M., Willson, R.C.: Solar Phys. **162**, 101–128 (1995)

43. Gabriel, A.H., Grec, G., Charra, J., Robillot, J.-M., Roca Cortés, T., Turck-Chiéze, S., Bocchia, R., Boumier, P., Cantin, M., Cespédes, E., Cougrand, B., Crétolle, J., Damé, L., Decaudin, M., Delache, P., Denis, N., Duc, R., Dzitko, H., Fossat, E., Fourmond, J.-J., García, R.A., Gough, D., Grivel, C., Herreros, J.M., Lagardére, H., Moalic, J.-P., Pallé, P.L., Pétrou, N., Sanchez, M., Ulrich, R., van der Raay, H.B.: Solar Phys. **162**, 61–99 (1995)

44. Gergely, T.E., Kundu, M.R.: Solar Phys. **34**, 433–446 (1974)

45. Gloeckler, G., Fisk, L.A., Hefti, S., Schwadron, N.A., Zurbuchen, T.H., Ipavich, F.M., Geiss, J., Bochsler, P., Wimmer-Schweingruber, R.F.: Geophys. Res. Lett. **26**, L157–L160 (1999)

46. Gold, T.: In: van de Hulst, J.C., Burgers, J.M. (eds.) Gas Dynamics of Cosmic Clouds, p. 103. North-Holland, New York (1955)

47. González-Esparza, A.: Space Sci. Rev. **97**, 197–200 (2001)

48. Gosling, J.T.: In: Russell, C.T., Priest, E.R., Lee, L.C. (eds.) Geophysical Monograph Series, vol. 58, p. 343. AGU, Washington, D.C. (1990)

49. Gosling, J.T.: J. Geophys. Res. **98**, 18937–18949 (1993)

50. Gosling, J.T.: Coronal Mass Ejections in the Solar Wind at High Solar Latitudes: An Overview. Presented at the Third SOHO Workshop, Estes Park (1994)

51. Gosling, J.T., Asbridge, J.R., Bames, S.J., Hundhausen, A.J., Strong, I.B.: J. Geophys. Res. **73**, 43–50 (1968)

52. Gosling, J.T., Hildner, E., MacQueen, R.M., Munro, R.H., Poland, A.I., Ross, C.L.: J. Geophys Res. **79**, 4581–4587 (1974)

53. Gosling, J.T., Hildner, E., MacQueen, R.M., Munro, R.H., Poland, A.I., Ross, C.L.: Solar Phys. **40**, 439–448 (1975)

54. Gosling, J.T., Hildner, E., MacQueen, R.M., Munro, R.H., Poland, A.I., Ross, C.L.: Solar Phys. **48**, 389–397 (1976)

55. Gosling, J.T., Bame, S.J., McComas, D.J., Phillips, J.L., Scime, E.E., Pizzo, V.J., Goldstein, B.E., Balogh, A.: Geophys. Res. Lett. **21**, L237–L240 (1994)

56. Gosling, J.T., Skoug, R.M., McComas, D.J., Smith, C.W.: J. Geophys. Res. **110**, A01107 (2005). doi:10.1029/20043A010809

57. Gosling, J.T., Skoug, R.M., McComas, D.J., Smith, C.W.: Geophys. Res. Lett. **32**, L05105 (2005). doi:10.1029/2005GL022406
58. Greaves, W.M.H, Newton, H.W.: Mon. Not. R. Astron. Soc. **88**, 556–567 (1928)
59. Greaves, W.M.H, Newton, H.W.: Mon. Not. R. Astron. Soc. **89**, 84–92 (1928)
60. Gloeckler, G., Geiss, J., Roelof, E.C., Fisk, L.A., Ipavich, F.M., Ogilvie, K.W., Lanzerotti, L.J., von Steiger, R., Wilken, B.: J. Geophys. Res. **99**, 17637–17643 (1994)
61. Kanekal, S.G., Baker, D.N., Blake, J.B., Klecker, B., Mewaldt, R.A., Mason, G.M.: J. Geophys. Res. **104**, 24885–24894 (1999)
62. Hale, G.E.: Astrophys. J. **73**, 379–412 (1931)
63. Handy, B.N., Acton, L.W., Kankelborg, C.C., Wolfson, C.J., Akin, D.J., Bruner, M.E., Caravalho, R., Catura, R.C., Chevalier, R., Duncan, D.W., Edwards, C.G., Feinstein, C.N., Freeland, S.L., Friedlander, F.M., Hoffmann, C.H., Hurlburt, N.E., Jurcevich, B.K., Katz, N.L., Kelly, G.A., Lemen, J.R., Levay, M., Lindgren, R.W., Mathur, D.P., Meyer, S.B., Morrison, S.J., Morrison, M.D., Nightingale, R.W., Pope, T.P., Rehse, R.A., Schrijver, C.J., Shine, R.A., Shing, L., Tarbell, T.D., Title, A.M., Torgerson, D.D., Golub, L., Bookbinder, J.A., Caldwell, D., Cheimets, P.N., Davis, W.N., DeLuca, E.E., McMullen, R.A., Amato, D., Fisher, R., Maldonado, H., Parkinson, C.: Solar Phys. **187**, 229–260 (1999)
64. Harrison, R.A.: Astron. Astrophys. **162**, 283–291 (1986)
65. Harrison, R.A.: Solar Phys. **166**, 441–444 (1996)
66. Harrison, R.A., Simnett, G.M.: Adv. Space Res. **4**, 199–202 (1984)
67. Harrison, R.A., Waggett, P.W., Bentley, R.D., Phillips, K.J.H., Bruner, M., Dryer, M., Simnett, G.M.: Solar Phys. **97**, 387–400 (1985)
68. Harrison, R.A., Sime, D.G.: J. Geophys. Res. **94**, 2333–2344 (1989)
69. Harrison, R.A., Hildner, E., Hundhausen, A.J., Sime, D.G., Simnett, G.M.: J. Geophys. Res. **95**, 917–937 (1990)
70. Harrison, R.A., Sawyer, E.C., Carter, M.K., Cruise, A.M., Cutler, R.M., Fludra, A., Hayes, R.W., Kent, B.J., Lang, J., Parker, D.J., Payne, J., Pike, C.D., Peskett, S.C., Richards, A.G., Gulhane, J.L., Norman, K., Breeveld, A.A., Breeveld, E.R., Al Janabi, K.F., McCalden, A.J., Parkinson, J.H., Self, D.G., Thomas, P.D., Poland, A.I., Thomas, R.J., Thompson, W.T., Kjeldseth-Moe, O., Brekke, P., Karud, J., Maltby, P., Aschenbach, B., Br uninger, H., Kúhne, M., Hollandt, J., Siegmund, O.H.W., Huber, M.C.E., Gabriel, A.H., Mason, H.E., Bromage, B.J.I.: Solar Phys. **162**, 233–290 (1995)
71. Henke, T., Woch, J., Mall, U., Livi, S., Wilkin, R., Schwenn, R., Gloeckler, G., von Steiger, R., Forsyth, R.J., Balogh, A.: Geophys. Res. Lett. **25**, L3465–L3468 (1998)
72. Henke, T., Woch, J., Schwenn, R., Mall, U., Gloeckler, G., von Steiger, R., Forsyth, R.J., Balogh, A.: J. Geophys. Res. **106**, 10597–10613 (2001)
73. Hetherington, B.: A Chronicle of Pre-telescopic Astronomy. Wiley, Chichester/New York (1996)
74. Hewish, A.: Solar Phys. **116**, 195–198 (1988)
75. Hewish, A., Bravo, S.: Solar Phys. **106**, 185–200 (1986)
76. Hewish, A., Symonds, M.D.: Planet. Space Sci. **17**, 313–320 (1969)
77. Hewish, A., Scott, P.F., Wills, D.: Nature **203**, 1214–1217 (1964)
78. Hick, P., Jackson, B.V.: Adv. Space Res. **14**, 135–138 (1994)
79. Hick, P., Jackson, B.V., Schwenn, R.: Astron. Astrophys. **244**, 242–250 (1991)
80. Hildner, E.: Proceedings of the AIAA Meeting. Washington, DC (1974)
81. Hirshberg, J., Bame, S.J., Robbins, D.E.: Solar Phys. **23**, 467–486 (1972)
82. Holman, G.D.: The RHESSI Homepage, Available via NASA/GSFC. http://hesperia.gsfc.nasa.gov/hessi/ (2008). Cited 10 Nov 2008
83. Houminer, Z.: Nat. Phys. Sci. **231**, 165–167 (1971)
84. Houminer, Z.: Planet. Space Sci. **21**, 1617–1624 (1973)
85. Houminer, Z., Hewish, A.: Planet. Space Sci. **20**, 1703–1716 (1972)
86. Houminer, Z., Hewish, A.: Planet. Space Sci. **22**, 1041–1042 (1974)
87. Houminer, Z., Hewish, A.: Planet. Space Sci. **36**, 301–306 (1988)

88. Hovestadt, D., Hilchenbach, M., Bürgi, A., Klecker, B., Laeverenz, P., Scholer, M., Grünwaldt, H., Axford, W.I., Livi, S., Marsch, E., Wilken, B., Winterhoff, H.P., Ipavich, F.M., Bedini, P., Coplan, M.A., Galvin, A.B., Gloeckler, G., Bochsler, P., Balsiger, H., Fischer, J., Geiss, J., Kallenbach, R., Wurz, P., Reiche, K.-U., Gliem, F., Judge, D.L., Ogawa, H.S., Hsieh, K.C., Möbius, E., Lee, M.A., Managadze, G.G., Verigin, M.I., Neugebauer, M.: Solar Phys. **162**, 441–481 (1995)
89. Howard, R.A., Koomen, M.J., Michels, D.J., Tousey, R., Dewiler, C.R., Roberts, D.E., Seal, R.T., Whitney, J.T., Hansen, R.T., Hansen, S.F., Garcia, C.J., Yasukawa, E.: World Data Center A, Report UAG 48A (1975)
90. Howard, R.A., Michels, D.J., Sheeley, N.R., Jr., Koomen, M.J.: Astrophys. J. **263**, L101–L104 (1982)
91. Howard, R.A., Sheeley, N.R., Jr., Michels, D.J., Koomen, M.J.: J. Geophys. Res. **90**, 8173–8191 (1985)
92. Howard, R.A., Moses, J.D., Vourlidas, A., Newmark, J.S., Socker, D.G., Plunkett, S.P., Korendyke, C.M., Cook, J.W., Hurley, A., Davila, J.M., Thompson, W.T., St Cyr, O.C., Mentzell, E., Mehalick, K., Lemen, J.R., Wuelser, J.P., Duncan, D.W., Tarbell, T.D., Wolfson, C.J., Moore, A., Harrison, R.A., Waltham, N.R., Lang, J., Davis, C.J., Eyles, C.J., Mapson-Menard, H., Simnett, G.M., Halain, J.P., Defise, J.M., Mazy, E., Rochus, P., Mercier, R., Ravet, M.F., Delmotte, F., Auchere, F., Delaboudiniere, J.P., Bothmer, V., Deutsch, W., Wang, D., Rich, N., Cooper, S., Stephens, V., Maahs, G., Baugh, R., McMullin, D., Carter, T.: Space Sci. Rev. **136**, 67–115 (2008)
93. Howard, T.: Coronal Mass Ejections, An Introduction. Springer, New York (2011)
94. Howard, T.A., Tappin, S.J.: Astron. Astrophys. **440**, 373–383 (2005)
95. Howard, T.A., Tappin, S.J.: Solar Phys. **252**, 373–383 (2008)
96. Howard, T.A., Tappin, S.J.: Space Sci. Rev. **147**, 31–54 (2009)
97. Howard, T.A., Tappin, S.J.: Space Sci. Rev. **147**, 89–110 (2009)
98. Howard, T.A., Tappin, S.J.: Space Weather **8**, S07004 (2010). doi:10.1029/2009SW000531
99. Howard, T.A., Webb, D.F., Tappin, S.J., Mizuno, D.R., Johnston, J.C.: J. Geophys. Res. **111** (2006). doi:10.1029/2005JA011349
100. Howard, T.A., Fry, C.D., Johnston, J.C., Webb, D.F.: Astrophys. J. **667**, 610–625 (2007)
101. Hudson, H.S.: Solar Phys. **113**, 1–9 (1987)
102. Hudson, H.S, Haisch, B., Strong, K.T.: J. Geophys. Res. **100**, 3473–3477 (1995)
103. Hundhausen, A.J.: In: Sonnet, C.P., Coleman, P.J., Wilcox, J.M. (eds.) Solar Wind, p. 393. NASA Special Publication, SP-308 (1972)
104. Hundhausen, A.J.: Coronal Expansion and Solar Wind. Springer, New York (1972)
105. Hundhausen, A.J.: In: Pizzo, V.J., Holzer, T.E., Sime, D.G. (eds.) Proceedings of 6th International Solar Wind Conference, Estes Park, p. 192. HAO (1987)
106. Hundhausen, A.J., Bame, S.J., Montgomery, M.D.: J. Geophys. Res. **75**, 4631–4642 (1970)
107. Hundhausen, A.J., Sawyer, C.B., House, L., Illing, R.M.E., Wagner, W.J.: J. Geophys. Res. **89**, 2639–2646, (1984)
108. Hundhausen, A.J., Burkepile, J.T., St. Cyr, O.C.: J. Geophys. Res. **99**, 6543–6552 (1994)
109. Hurlburt, N.: The TRACE Webpage, Available via Lockheed Martin Missile and Space. http://trace.lmsal.com/ (2000). Cited June 2000
110. Ichimoto, K., The Solar-B Team: J. Korean Astron. Soc. **38**, 307–310 (2005)
111. Illing, R.M.E., Hundhausen, A.J.: J. Geophys. Res. **90**, 275–282 (1985)
112. Ivanov, K.G.: Astronomicheskii Zhurnal **48**, 998 (1971)
113. Ivanov, K.G.: Sov. Astron. **17**, 94 (1973)
114. Jackson, B.V.: Solar Phys. **95**, 363–370 (1985)
115. Jackson, B.V.: Adv. Space Res. **6**, 307–310 (1986)
116. Jackson, B.V.: Adv. Space Res. **9**, 69–74 (1989)
117. Jackson, B.V., Froehling, H.R.: Astron. Astrophys. **299**, 885–892 (1995)
118. Jackson, B.V., Hick, P.P.: Solar Phys. **211**, 345–356 (2002)
119. Jackson, B.V., Leinert, C.: J. Geophys. Res. **90**, 10759–10764 (1985)

120. Jackson, B.V., Howard, R.A., Sheeley, N.R., Jr., Michels, D.J., Koomen, M.J., Illing, R.M.E.:
     J. Geophys. Res. **90**, 5075–5081 (1985)
121. Jackson, B.V., Rompolt, B., Švestka, Z.: Solar Phys. **115**, 327–343 (1988)
122. Jackson, B.V., Buffington, A., Hick, P.P., Altrock, R.C., Figueroa, S., Holladay, P.E., Johnston,
     J.C., Kahler, S.W., Mozer, J.B., Price, S., Radick, R.R., Sagalyn, R., Sinclair, D., Simnett,
     G.M., Eyles, C.J., Cooke, M.P., Tappin, S.J., Kuchar, T., Mizuno, D., Webb, D.F., Anderson,
     P.A., Keil, S.L., Gold, R.E., Waltham, N.R.: Solar Phys. **225**, 177–207 (2004)
123. Jackson, B.V., Buffington, A., Hick, P.P., Wang, X., Webb, D.F.: J. Geophys. Res. **111** (2006).
     doi:10.1029/2004JA010942
124. Janardhan, P., Balasubramanian, V., Ananthakrishnan, S.: In: Wilson, A. (ed.) Proceedings of
     the 31st ESLAB Symposium, Noordwijk, p. 177. ESA SP-415 (1997)
125. Jones, R., Canals, A., Breen, A., Fallows, R., Lawrence, G.: Proceedings of the 1st EGU
     Science Assembly, Nice. EGU-A-00452 (2004)
126. Joselyn, J.A., McIntosh, P.S.: J. Geophys. Res. **86**, 4555–4564 (1981)
127. Kahler, S.W.: Ann. Rev. Astron. Astrophys. **30**, 113–141 (1992)
128. Kahler, S.W., Reames, D.V. Sheeley, N.R., Jr.: Astrophys. J. **562**, 558–565 (2001)
129. Kaiser, M.L., Kucera, T.A., Davila, J.M., St. Cyr, O.C., Guhathakurta, M., Christian, E.: Space
     Sci. Rev. **136**, 5–16 (2008)
130. Klein, L.W., Burlaga, L.F.: J. Geophys. Res. **87**, 613–624 (1982)
131. Kohl, J.L., Esser, R., Gardner, L.D., Habbal, S., Daigneau, P.S., Dennis, E.F., Nystrom, G.U.,
     Panasyuk, A., Raymond, J.C., Smith, P.L., Strachan, L., van Ballegooijen, A.A., Noci, G.,
     Fineschi, S., Romoli, M., Ciaravella, A., Modigliani, A., Huber, M.C.E., Antonucci, E.,
     Benna, C., Giordano, S., Tondello, G., Nicolosi, P., Naletto, G., Pernechele, C., Spadaro,
     D., Poletto, G., Livi, S., von der Lühe, O., Geiss, J., Timothy, J.G., Gloeckler, G., Allegra,
     A., Basile, G., Brusa, R., Wood, B., Siegmund, O.H.W., Fowler, W., Fisher, R., Jhabvala, M.:
     Solar Phys. **162**, 313–356 (1995)
132. Koomen, M., Howard, R.A., Hansen, R., Hansen, S.: Solar Phys. **34**, 447–452 (1974)
133. Korreck, K.E., Zurbuchen, T.H., Lepri, S.T., Raines, J.M.: Astrophys. J. **659**, 773–779 (2007)
134. Lazarus, A.J., Binsack, J.H.: Ann. I. Q. S. Y. **3**, 378–385 (1969)
135. Lazarus, A.J., Ogilvie, K.W., Burlaga, L.F.: Solar Phys. **13**, 232–239 (1970)
136. Lazarus, A.J., Richardson, J.D., Decker, R.B., McDonald, F.B.: Space Sci. Rev. **89**, 53–59
     (1999)
137. Leamon, R.J., Canfield, R.C., Jones, S.L., Lambkin, K., Lundberg, B.J., Pevtsov, A.A.: J.
     Geophys. Res. **109** (2004). doi:10.1029/2003JA010324
138. Leinert, C., Link, H., Pitz, E., Salm, N., Kluppelberg, D.: Raumfahrtforschung **19**, 264–267
     (1975)
139. Lepping, R.P., Jones, J.A., Burlaga, L.F.: J. Geophys. Res. **95**, 11957–11965 (1990)
140. Lepping, R.P., Wu, C.-C., McClernan, K.: J. Geophys. Res. **108** (2003).
     doi:10.1029/2002JA009640
141. Lepri, S.T., Zurbuchen, T.H., Fisk, L.A., Richardson, I.G., Cane, H.V., Gloeckler, G.: J.
     Geophys. Res. **106**, 29231–29238 (2001)
142. Liewer, P.C., De Jong, E.M., Hall, J.R., Howard, R.A., Thompson, W.T., Culhane, J.L., Bone,
     L., van Driel-Gesztelyi, L.: Solar Phys. **256**, 57–72 (2009)
143. Lin, R.P., Dennis, B.R., Hurford, G.J., Smith, D.M., Zehnder, A., Harvey, P.R., Curtis, D.W.,
     Pankow, D., Turin, P., Bester, M., Csillaghy, A., Lewis, M., Madden, N., van Beek, H.F.,
     Appleby, M., Raudorf, T., McTiernan, J., Ramaty, R., Schmahl, E., Schwartz, R., Krucker,
     S., Abiad, R., Quinn, T., Berg, P., Hashii, M., Sterling, R., Jackson, R., Pratt, R., Campbell,
     R.D., Malone, D., Landis, D., Barrington-Leigh, C.P., Slassi-Sennou, S., Cork, C., Clark,
     D., Amato, D., Orwig, L., Boyle, R., Banks, I.S., Shirey, K., Tolbert, A.K., Zarro, D.,
     Snow, F., Thomsen, K., Henneck, R., McHedlishvili, A., Ming, P., Fivian, M., Jordan, John,
     Wanner, Richard, Crubb, Jerry, Preble, J., Matranga, M., Benz, A., Hudson, H., Canfield, R.C.,
     Holman, G.D., Crannell, C., Kosugi, T., Emslie, A.G., Vilmer, N., Brown, J.C., Johns-Krull,
     C., Aschwanden, M., Metcalf, T., Conway, A.: Solar Phys. **210**, 3–32 (2002)
144. Liu, Y., Richardson, J.D., Belcher, J.W.: Planet. Space Sci. **53**, 3–17 (2005)

145. Lynch, B.J., Zurbuchen, T.H., Fisk, L.A., Antiochos, S.K.: J. Geophys. Res. **108** (2003). doi:10.1029/2002JA009591
146. Lynch, B.J., Gruesbeck, J.R., Zurbuchen, T.H.: J. Geophys. Res. **108** (2005). doi:10.1029/2005JA011137
147. Lyot, M.B.: Mon. Not. R. Astron. Soc. **99**, 578–590 (1939)
148. MacQueen, R.M., Csoeke-Poeckh, A., Hildner, E., House, L., Reynolds, R., Stanger, A., Tepoel, H., Wagner, W.: Solar Phys. **65**, 91–107 (1980)
149. MacQueen, R.M., Burkepile, J.T., Holzer, T.E., Stanger, A.L., Spence, K.E.: Astrophys. J. **549**, 1175–1182 (2001)
150. Marubashi, K.: In: Crooker, N., Joselyn, J.A., Feynman, J. (eds.) Geophysical Monograph Series, vol. 99, p. 147. AGU, Washington, D.C. (1997)
151. Maunder, E.W.: Mon. Not. R. Astron. Soc. **65**, 2–34 (1904)
152. McKenzie, D.: Yohkoh Solar Observatory Webpage, Available via Montana State University. http://solar.physics.montana.edu/sxt/ (2006). Cited 29 Nov 2006
153. McLean, D.J., Labrum, N.R. (eds.): Solar Radiophysics: Studies of Emission from the Sun at Metre Wavelengths. Cambridge University Press, Cambridge/New York (1985)
154. Meyer, P., Parker, E.N., Simpson, J.A.: Phys. Rev. **104**, 768–783 (1956)
155. Michels, D.J., Howard, R.A., Koomen, M.J., Sheeley, N.R., Jr.: In: Kundu, M.R., Gergely, T.E. (eds.) Radio Physics of the Sun, p. 439. D. Reidel, Hingham (1980)
156. Michels, D.J., Sheeley, N.R., Jr., Howard, R.A., Koomen, M.J., Schwenn, R., Mulhauser, K.H., Rosenbauer, H.: Adv. Space Res. **4**, 311–321 (1984)
157. Mierla, M., Davila, J., Thompson, W., Inhester, B., Srivastava, N., Kramar, M., St. Cyr, O.C., Stenborg, G., Howard, R.A.: Solar Phys. **252**, 385–396 (2008)
158. Mierla, M., Inhester, B., Antunes, A., Boursier, Y., Byrne, J.P., Colaninno, R., Davila, J., de Koning, C.A., Gallagher, P.T., Gissot, S., Howard, R.A., Howard, T.A., Kramar, M., Lamy, P., Liewer, P.C., Maloney, S., Marqué, C., McAteer, R.T.J., Moran, T., Rodriguez, L., Srivastava, N., St. Cyr, O.C., Stenborg, G., Temmer, M., Thernisien, A., Vourlidas, A., West, M.J., Wood, B.E., Zhukov, A.N.: Ann. Geophys. **28**, 203–215 (2010)
159. Morrison, P.: Phys. Rev. **95**, 646 (1954)
160. Moullard, O., Burgess, D., Salem, C., Mangeney, A., Larson, D.E., Bale, S.D.: J. Geophys. Res. **106**, 8301–8314 (2001)
161. Müller-Mellin, R., Kunow, H., Fleißner, V., Pehlke, E., Rode, E., Röschmann, N., Scharmberg, C., Sierks, H., Rusznyak, P., McKenna-Lawlor, S., Elendt, I., Sequeiros, J., Meziat, D., Sanchez, S., Medina, J., Del Peral, L., Witte, M., Marsden, R., Henrion, J.: Solar Phys. **162**, 483–504 (1995)
162. Munro, R.H., Sime, D.G.: Solar Phys. **97**, 191–201 (1985)
163. Newton, H.W.: Observatory **62**, 317–326 (1939)
164. Newton, H.W.: Mon. Not. R. Astron. Soc. **103**, 244–257 (1943)
165. Ogawara, Y.: Solar Phys. **113**, 361–370 (1987)
166. Ogawara, Y., Takano, T., Kato, T., Kosugi, T., Tsuneta, S., Wtanabe, T., Kondo, I., Uchida, Y.: Solar Phys. **136**, 1–16 (1991)
167. Ogilvie, K.W., Burlaga, L.F.: Solar Phys. **8**, 422–434 (1969)
168. Ogilvie, K.W., von Rosenvinge, T.T, Durney, A.C.: Science **198**, 131–138 (1977)
169. Ogilvie, K.W., Durney, A.C., von Rosenvinge, T.T.: IEEE Trans. Geosci. Electron. **GE-16**, 151–153 (1978)
170. Papitashvili, N.E.: IMP-8 Project Information, Available via NASA/GSFC. http://spdf.gsfc.nasa.gov/imp8/project.html (2000). Cited 2000
171. Pesnell, D, Addison, K.: The Solar Dynamics Observatory Webpage, Available via NASA/GSFC. http://sdo.gsfc.nasa.gov/ (2009). Cited 2 Apr 2009
172. Pudovkin, M.I.: J. Geophys. Res. **100**, 7917–7919 (1995)
173. Qiu, J., Hu, Q., Howard, T.A., Yurchyshyn, V.B.: Astrophys. J. **659**, 758–772 (2007)
174. Ranyard, C.A.: Mem. R. Astron. Soc. **41**, 520 (1879)
175. Reid, G.C., Leinbach, H.: J. Geophys. Res. **64**, 1801–1805 (1959)
176. Reinard, A.A.: Astrophys. J. **682**, 1289–1305 (2008)

177. Richardson, I.G., Cane, H.V.: J. Geophys. Res. **109** (2004). doi: 10.1029/2004JA010598
178. Richardson, J.D., Paularena, K.I., Wang, C., Burlaga, L.F.: J. Geophys. Res. **107** (2002). doi:10.1029/2001JA000175
179. Richter, I., Leinert, C., Planc, B.: Astron. Astrophys. **110**, 115–120 (1982)
180. Rickett, B.J.: Solar Phys. **43**, 237–247 (1975)
181. Riley, P., Linker, J.A., Lionello, R., Mikič, Odstrcil, D., Hidalgo, M.A., Cid, C., Hu, Q., Lepping, R.P., Lynch, B.J., Rees, A.: J. Atmos. Solar Terr. Phys. **66**, 1321–1331 (2004)
182. Rodríguez-pacheco, J., Cid, C., Blanco, J.J., Sequeiros, J.: Solar Phys. **213**, 121–145 (2003)
183. Sabine, E.: Philos. Trans. R. Soc. Lond. **142**, 103–124 (1852)
184. Sanderson, T.R., Marsden, R.G., Heras, A.M., Wenzel, K.-P., Anglin, J.D., Balogh, A., Forsyth, R.: Geophys. Res. Lett. **19**, L1263–L1266 (1992)
185. Scherrer, P.H., Bogart, R.S., Bush, R.I., Hoeksema, J.T., Kosovichev, A.G., Schou, J., Rosenberg, W., Springer, L., Tarbell, T.D., Title, A., Wolfson, C.J., Zayer, I., MDI Engineering Team: Solar Phys. **162**, 129–188 (1995)
186. Schröder, W., Wiederkehr, K-H.: J. Atmos. Solar Terr. Phys. **63**, 1649–1660 (2001)
187. Sime, D.G., Hundhausen, A.J.: J. Geophys. Res. **92**, 1049–1055 (1987)
188. Sime, D.G., MacQueen, R.M., Hundhausen, A.J.: J. Geophys. Res. **89**, 2113–2121 (1984)
189. Simnett, G.M., Harrison, R.A.: Adv. Space Res. **4**, 279–282 (1984)
190. Simnett, G.M., Harrison, R.A.: Solar Phys. **99**, 291–311 (1985)
191. Sisko, G., Crooker, N.U., Clauer, C.R.: Adv. Space Res. **38**, 173–179 (2006)
192. Sonnet, C.P., Colburn, D.S., Davis, L., Smith, E.J., Coleman, P.J.: Phys. Rev. Lett. **13**, 153–156 (1964)
193. Stanger, A.L.: SMM C/P CME Event: 14 April 1980—Day of Year (DOY): 105 Available via HAO. http://smm.hao.ucar.edu/smm/smmcp_events/1980apr14.html (2000). Cited 29 Aug 2000
194. Švestka, Z.F.: Solar Phys. **160**, 53–56 (1995)
195. Tappin, S.J.: PhD Thesis, Transient disturbances in the solar wind. University of Cambridge (1984)
196. Tappin, S.J.: Solar Phys. **233**, 233–248 (2006)
197. Tappin, S.J., Howard, T.A.: Space Sci. Rev. **147**, 55–87 (2009)
198. Tappin, S.J., Hewish, A., Gapper, G.R.: Planet. Space Sci. **31**, 1171–1176 (1983)
199. Tappin, S.J., Buffington, A., Cooke, M.P., Eyles, C.J., Hick, P.P., Holladay, P.E., Jackson, B.V., Johnston, J.C., Kuchar, T., Mizuno, D., Mozer, J.B., Price, S., Radick, R.R., Simnett, G.M., Sinclair, D., Waltham, N.R., Webb, D.F.: Geophys. Res. Lett. **31** (2004). doi:10.1029/2003GL018766
200. Tappin, S.J., Howard, T.A., Hampson, M.M., Thompson, R.N., Burns, C.E.: J. Geophys. Res. **117**, A05103 (2012). doi:10.1029/2011JA017439
201. Taylor, H.E.: Solar Phys. **6**, 320–334 (1969)
202. Torsti, J., Valtonen, E., Lumme, M., Peltonen, P., Eronen, T., Louhola, M., Riihonen, E., Schultz, G., Teittinen, M., Ahola, K., Holmlund, C., Kelhä, V., Leppälä, K., Ruuska, P., Strömmer, E.: Solar Phys. **162**, 505–531 (1995)
203. Tousey, R.: In: Rycroft, M.J., Runcorn, S.K. (eds.) Space Research XIII, pp. 713–730. Akademie, Berlin (1973)
204. Trella, M., Greenfield, M., Herring, E.L., Credland, J., Freeman, H.R., Laine, R., Kilpatrick, W., Machi, D., Reth, A., Smith, A.: SOHO Mission Interruption Join NASA/ESA Investigation Board: Final Report, Available via ESA. http://sohowww.estec.esa.nl/whatsnew/SOHO_final_report.html (1998). Cited 31 Aug 1998
205. Tsurutani, B.T., Gonzalez, W.D., Lakhina, G.S., Alex, S.: J. Geophys. Res. **108**, 1268 (2002). doi:10.1029/2002JA009504
206. Vlasov, V.I.: Geomagn. Aeron. **21**, 324–326 (1981)
207. Vlasov, V.I.: Geomagn. Aeron. **22**, 446–450 (1982)
208. Vlasov, V.I., Shishov, V.I., Shisova, T.D.: Geomagn. Aeron. **25**, 211–214 (1985)
209. Wang, C., Richardson, J.D.: Geophys. Res. Lett. **29** (2002). doi:10.1029/2001GL014472
210. Watanabe, T.: Proceedings of the COSPAR, Plenary Meeting (1977)

211. Watanabe, T., Kakinuma, T., Kojima, M., Shibasaki, K.: J. Geophys. Res. **78**, 8364–8366 (1973)
212. Webb, D.F., Jackson, B.V.: J. Geophys. Res. **95**, 20641–20661 (1990)
213. Webb, D.F., Cheng, C.-C., Dulk, G.A., Edberg, S.J., Martin, S.F., McKenna-Lawlor, S., McLean, D.J.: In: Sturrock, P.A. (ed.) Solar Flares: A Monograph from Skylab Workshop II, p. 471. Colorado Associated University Press, Boulder (1980)
214. Webb, D.F., Jackson, B.V., Hick, P., Schwenn, R., Bothmer, V., Reames, D.: Adv. Space Res. **13**, 71–74 (1993)
215. Webb, D.F., Mizuno, D.R., Buffington, A., Cooke, M.P., Eyles, C.J., Fry, C.D., Gentile, L.C., Hick, P.P., Holladay, P.E., Howard, T.A., Hewitt, J.G., Jackson, B.V., Johnston, J.C., Kuchar, T.A., Mozer, J.B., Price, S., Radick, R.R., Simnett, G.M., Tappin, S.J.: J. Geophys. Res. **111** (2006). doi:10.1029/2006JA011655
216. Webb, D.F., Howard, T.A., Fry, C.D., Kuchar, T.A., Odstrcil, D., Jackson, B.V., Bisi, M.M., Harrison, R.A., Morrill, J.S., Howard, R.A., Johnston, J.C.: Solar Phys. **256**, 239–269 (2009)
217. Wild, J.P., McCready, L.L.: Aust. J. Sci. Res. **3**, 387–398 (1950)
218. Wild, J.P., Murray, J.D., Rowe, W.C.: Aust. J. Phys. **7**, 439–459 (1954)
219. Wild, J.P., Roberts, J.A., Murray, J.D.: Nature **173**, 532–534 (1954)
220. Wild, J.P., Sheridan, K.V., Trent, G.H.: In: Bracewell, R.N. (ed.) Proceedings of the Paris Symposium Radio Astronomy, IAU/URSI, p. 176. Stanford University Press, Stanford (1959)
221. Wild, J.P., Smerd, S.F., Weiss, A.A.: Ann. Rev. Astron. Astrophys. **1**, 291–366 (1963)
222. Wilhelm, K., Curdt, W., Marsch, E., Schühle, U., Lemaire, P., Gabriel, A., Vial, J.-C., Grewing, M., Huber, M.C.E., Jordan, S.D., Poland, A.I., Thomas, R.J., Kühne, M., Timothy, J.G., Hassler, D.M., Siegmund, O.H.W.: Solar Phys. **162**, 189–231 (1995)
223. Wilson, R.M., Hildner, E.: Solar Phys. **91**, 169–180 (1984)
224. Woo, R., Armstrong, J.W.: Nature **292**, 608–610 (1981)
225. Woo, R., Armstrong, J.W.: Nature **304**, 756 (1983)
226. Woo, R., Armstrong, J.W., Sheeley, N.R., Jr., Howard, R.A., Michels, D.J., Koomen, M.J., Schwenn, R.: J. Geophys. Res. **90**, 154–162 (1985)
227. Zhang, G., Burlaga, L.F.: J. Geophys. Res. **93**, 2511–2518 (1988)
228. Zhao, X.-P.: J. Geophys. Res. **97**, 15051–15055 (1992)
229. Zirker, J.B.: Total Eclipses of the Sun. Princeton University Press, Princeton (1995)

Chapter 3
Observation and Modeling

We obtain information about CMEs via two primary means: measurement and mathematical modeling. Through modeling we can try out different physical scenarios to describe how CMEs are launched and evolve through the heliosphere, and through observation we can extract physical information about CMEs and their environment and compare with that predicted by modeling. Therefore as our analysis of CME datasets becomes more sophisticated we can inform the models with more precise information, thereby improving their accuracy and increasing our knowledge of the physics governing CMEs. In this chapter we review the means by which we observe and measure CMEs and discuss the models describing their onset and evolution.

3.1 Observing and Measuring CMEs

Coronal mass ejections are observed in visible white light by coronagraphs that block out the light from the photosphere. This is made possible because CMEs are comprised of plasma, so they contain large numbers of free electrons. The white light we observe is originally from the photosphere, and is scattered off these electrons in the corona and solar wind via a process called Thomson scattering. Wide-angle coronagraphs are broadband imagers, partly because this maximizes our ability to observe them and partly because spectral information begins to be lost in the high corona [72].

In order to completely understand the evolution of CMEs it is necessary to understand the meaning of changes in their appearance in coronagraph and heliospheric images. Are these due to physical changes to the CME itself, or do they reflect changes in the observed intensity of the CME due to the Thomson scattering physics? Do we continue to observe the same component of a CME as it moves through the sky, or do different parts of the structure become more detectable as a result of the scattering? How do projection effects come into play? Can we use the CME image and our knowledge of how the light reaches us to

T. Howard, *Space Weather and Coronal Mass Ejections*, SpringerBriefs in Astronomy, DOI 10.1007/978-1-4614-7975-8_3, © Timothy Howard 2014

reproduce physical characteristics of the CME itself (such as mass, trajectory, three-dimensional structure)? With careful analysis of the white light images and a detailed understanding of the Thomson scattering physics and projection geometry, it is possible to extract a great deal of information about CMEs from their white light images.

3.1.1 The Detection of Thomson Scattered Light

Much of the theory discussed in this section arises from two recent publications by the author and coworkers, specifically from Howard and Tappin [28] and Howard and DeForest [27]. A more detailed treatment of the physics of Thomson scattering appears in the author's introductory book on CMEs [26], and the reader may like to review the historical work that developed this theory. The author recommends the works of Schuster [59], Minnaert [51], Billings [4] and Jackson [34].

Thomson scattering is a special case of the general theory of the scattering of electromagnetic radiation by charged particles. If electromagnetic radiation is incident on a free particle carrying a charge, the particle will be accelerated. As it accelerates, it too will emit radiation, and in Thomson scattering the momentum transfer from the photon to the electron is ignored, so the frequency of the scattered light is the same as the incident. This is the (Thomson) scattered radiation which is emitted in all directions from the scattering particle. The pattern of scattered light is symmetrical about the direction of the incident wave, since the acceleration of the electron is confined to the plane perpendicular to the direction of its propagation. If an observer lies at a scattering angle (χ) of $0°$ or $180°$, then the scattered light from the electron will be unpolarized because the electron will appear to be displaced equally in all directions. If, however, the observer lies at $\chi = 90°$, then the electron will appear to move only in a linear motion perpendicular to the incident light and the observer will see linearly polarized scattered light, but the maximum magnitude of the electric field will be the same.

The intensity I of the scattered light is described in terms of two components: the component transverse to the radial component I_T and the radial component I_R. These are given by

$$I_T = \frac{\pi \sigma_e}{2z^2} \int_{\cos \Omega}^{1} I(1 + \cos^2 \omega) \mathrm{d}(\cos \omega), \tag{3.1}$$

and

$$I_R = \frac{\pi \sigma_e}{2z^2} \int_{\cos \Omega}^{1} I[(1 + \sin^2 \chi) + \cos^2 \omega (1 - 3 \sin^2 \chi)] \mathrm{d}(\cos \omega), \tag{3.2}$$

where σ_e is the differential cross-section for perpendicular scattering, z is the distance between the scattering point and the observer, Ω is the angular size

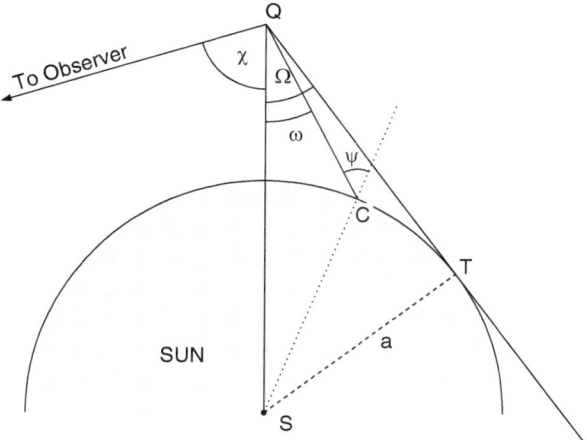

Fig. 3.1 The Sun and a nearby scattering point Q (modified from Fig. 6.2 of Billings [4]). The *shaded grey circle* represents the Sun with its center at S and radius *a*, T is the point where the scattered point vector crosses the Sun at a tangent and Ω is the angle between the tangent from Q and the SQ vector. The line of sight (To Observer) has been added and the angle between it and the SQ vector is shown as χ. The geometry of a ray from a point C is also shown, including the angles ψ and ω [28] (Reproduced with kind permission of Springer Science and Business Media)

of the solar disk relative to the scattering point, and χ is the scattering angle. A complete derivation of these equations can be found in Howard [26] and a diagram of the geometry used in the derivation can be found in Fig. 3.1. For simplicity we work with a component of the scattered light that allows the χ component to be factored out [26], so instead of I_R we use the polarized component $I_P = I_T - I_R$, or

$$I_P = -\frac{\pi \sigma_e}{2z^2} \int_{\cos \Omega}^{1} I \sin^2 \chi (1 - 3\cos^2 \omega) \mathrm{d}(\cos \omega). \qquad (3.3)$$

Using the following equation for solar limb darkening:

$$I = I_0 (1 - u + u \cos \psi), \qquad (3.4)$$

u is the limb darkening coefficient and ψ is the angle between the emitted radiation and radius vector, and applying some geometry Equations (3.1) and (3.3) become

$$I_T = I_0 \frac{\pi \sigma_e}{2z^2} [(1 - u)C + uD] \qquad (3.5)$$

and

$$I_P = I_0 \frac{\pi \sigma_e}{2z^2} \sin^2 \chi [(1 - u)A + uB]. \qquad (3.6)$$

The integrals within these equations have been assigned coefficients A, B, C and D, which are integrated to give

$$A = \cos\Omega\sin^2\Omega, \tag{3.7}$$

$$B = -\frac{1}{8}\left[1 - 3\sin^2\Omega - \frac{\cos^2\Omega}{\sin\Omega}(1 + 3\sin^2\Omega)\ln\left(\frac{1+\sin\Omega}{\cos\Omega}\right)\right], \tag{3.8}$$

$$C = \frac{4}{3} - \cos\Omega - \frac{\cos^3\Omega}{3}, \tag{3.9}$$

$$D = \frac{1}{8}\left[5 + \sin^2\Omega - \frac{\cos^2\Omega}{\sin\Omega}(5 - \sin^2\Omega)\ln\left(\frac{1+\sin\Omega}{\cos\Omega}\right)\right]. \tag{3.10}$$

The total intensity of the light scattered from the electrons is

$$I_{tot} = (I_T + I_R) = 2I_T - I_P, \tag{3.11}$$

where I_0 is the intensity of the Sun as a power per unit photospheric area per unit solid angle. The coefficients A, B, C and D are generally known as the van de Hulst coefficients after the author who modified the coefficients in order to reduce the number of tabulated functions needed [71]. The terms were actually originally introduced by Minnaert [51], and in the modern computer era the modifications imposed by van de Hulst are no longer required.

3.1.2   At Larger Angles from the Sun

The equations provided in the previous section apply when the scattering point is relatively close to the Sun (see Fig. 3.1). As the scattering point moves away from the Sun, the geometry becomes simplified since the Sun tends to appear as a point source. Notice that the van de Hulst coefficients are dependent on one variable only: the relative angular size of the Sun Ω. As the distance from the scattering point R increases this parameter decreases in size and all four of the coefficients tend towards $1/R^2$. This occurs early, when the scattering point is still only a few solar radii from the Sun (see Fig. 4 of Howard and Tappin [28] or Fig. 4.5 of Howard [26]).

Another important property is the spreading of the Thomson scattered signal. The scattered intensity is proportional to both the incident light and the density of the scattering volume. We know that the density of material and the incident light both decrease at an increasing rate of around R^2 as we move away from the Sun, so therefore the point that will have the brightest scattered light will be that which is closest to the Sun. Along any line of sight (a line of sight, LOS, is a vector from the observer through the point being measured) the point that is closest to the Sun is always the which is perpendicular to the solar radius vector. The locus of all such points along all lines of sight forms a sphere with the diameter as the Sun-observer line. Vourlidas and Howard [73] termed this the "Thomson surface" and stated that the detectability of features in the heliosphere was heavily dependent on the proximity of the feature to this sphere. The picture is complicated, however, by the fact that the third factor that governs the physics of the scattered light, that

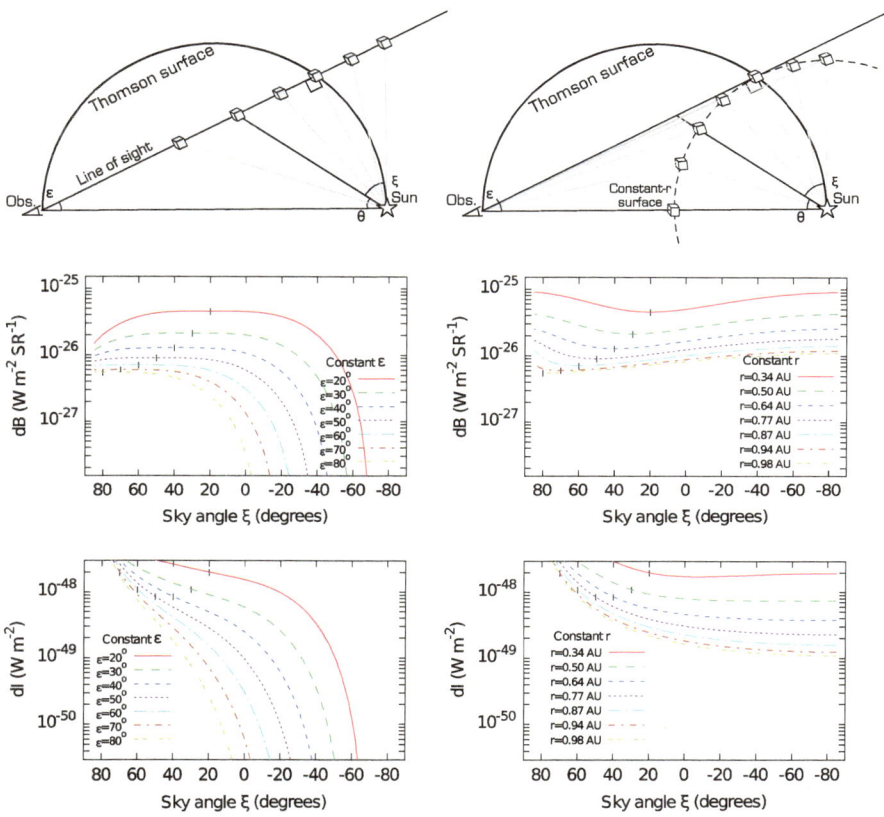

Fig. 3.2 Comparing the same hypothetical cube of plasma at constant r (rather than constant ε) eliminates illumination effects. *Left*: surface brightness dB and intensity dI at constant ε are important for understanding how to interpret individual images. *Right*: dB and dI at constant r show how well an instrument will detect CMEs at a given exit angle. In all plots, the intersection of each line with the Thomson surface is indicated with the *vertical bars*. Each parameter is plotted as a function of the angle from the sky, the Sky Angle ξ (From Howard and DeForest [27] and reproduced by permission of the AAS)

of the scattering efficiency itself, is actually *minimized* on this Thomson surface. This results in a broadening of the scattered intensity function meaning that the transient can actually be some considerable distance from the Thomson surface and still be detectable by a heliospheric imager. This has been proven with *STEREO*, which detected far-sided CMEs when they were on opposite sides of the Sun in their orbit [44]. Figure 3.2 shows the variation of surface brightness B and intensity I for a small volume of plasma held at a constant elongation ε and a constant radius r from the Sun.

3.1.3 Real Measurements

The theory discussed in the previous section describes the fundamental physics by which light from the CME reaches our detectors. In reality, there is much more to govern the detectability than simply the nature of the scattered light from the electrons within the CME. We must remember that it is not only the CME, but also the solar wind that contains free electrons and that it is all subjected to Thomson scattering. Our ability to detect CMEs, therefore, is not only governed by the properties of the CME itself, but also by the properties of the surrounding solar wind.

3.1.3.1 Integrated Lines of Sight

When one observes an object in space, one does not observe the object exclusively, but rather everything between the object and the observer is observed together collectively. The result is a two-dimensional projection representing an integration of everything along the direction observed. A single direction along which one observes is called a line of sight, and the resulting two-dimensional projection of everything on that line is called an integrated line of sight. When observing CMEs we are presented with the same problem of integrated lines of sight, except here it is far more difficult to identify depth in the resulting image. The theory of Thomson scattering regarding CMEs is presented as a line-of-sight integral, but this is not strictly true. It is in fact an integral though the cone of the observer's (instrument's) point spread function. In other words, it is not restricted to a single line integral, but rather is a function of the collecting area and beam size of the instrument. The total intensity integrated along the LOS received by the detector [28] is

$$I = \int_0^\infty N_e z^2 G \, \mathrm{d}z, \tag{3.12}$$

where G is the scattering expression given by

$$G_T = \frac{\pi \sigma_e}{2z^2}[(1-u)C + uD],$$
$$G_P = \frac{\pi \sigma_e}{2z^2}[(1-u)A + uB],$$
$$G_R = G_T - G_P.$$

When observing CMEs, it is generally assumed that the entire intensity is located on the Thomson surface, so the integral in Eq. (3.12) is reduced to

$$I_{\mathrm{rec}} = z_T^2 N_e G, \tag{3.13}$$

where z_T is the distance from the observer to the Thomson surface and N_e and G are both functions of z_T (i.e., $N_e = N_e(z_T)$ and $G = G(z_T)$). The spreading of the signal created by the partial canceling by the minimizing of the scattering efficiency results in this assumption being applicable even across fairly large distances from the Thomson surface. This allows a gain in accuracy in the determination of parameters such as density, but a loss of information about the location of the scattering volume itself. Howard and DeForest [27] describe the means by which CME density and mass calculation vary across different locations in the heliosphere.

3.1.3.2 Signal-to-Noise

The detectability of a feature is determined by its level above a noise floor [27]. In the corona there is a strong elongation dependence on the background noise because coronal images have very sharp gradients in intensity [5]. In the solar wind, much of the background light arises from uniform random variables that are independently sampled in each resolution element of an image. Assuming the background noise is a random variable of uniform characteristics and a constant number of samples per unit solid angle, the noise against which a feature is detected scales as $N = L\Phi$, where L is a coefficient dependent on the instrument and background subtraction level and Φ is the apparent size of the feature being measured. The signal-to-noise S/N ratio of a given detection thus scales as

$$S/N = B\Phi^{0.5}/L, \tag{3.14}$$

where B is the radiance measured as an average pixel value. Figure 3.3 shows the variation of S/N with the sky angle for features observed at large distances from the Sun (i.e., by heliospheric imagers). An interesting result that arises from these plots is that measurements at a constant elongation ε do not peak on the Thomson surface itself, but rather at a location closer to the observer (the sky angle is measured relative to the plane of the sky and is positive in the direction towards the observer).

3.1.3.3 Structure and Location

The final factors to consider are the properties of the observed feature itself. When close to the Sun (i.e., observed by coronagraphs), much of the information about the structure and 3-D location of the CME is lost because the narrow field of view of coronagraphs provides only relatively small features, the structure of which do not play a significant roll in its appearance. We can extract more information about the 3-D structure and location of feature by observing them from multiple angles; such has been possible with *STEREO* [50], but efforts to perform this with any degree of accuracy have been met with only mixed success [31]. Close to the Sun the following two assumptions are applied:

- Every measured point lies in the plane of the sky, so $r = \sin \varepsilon$. This is called the Point-P approximation [25].
- The elongation angle ε is small, so $\sin \varepsilon \sim \varepsilon$.

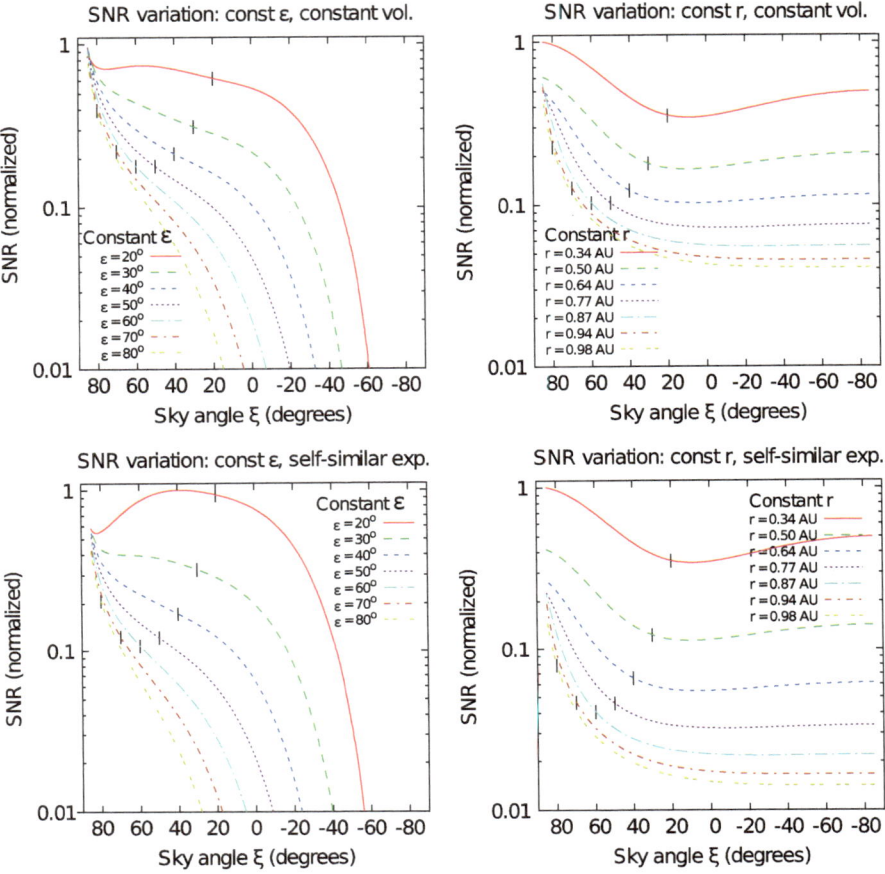

Fig. 3.3 Variation of signal-to-noise (S/N) for hypothetical features with a constant-brightness background, for features observed at large distances from the Sun under various conditions [27]. Reproduced with permission of the AAS. The *left column* shows variation along the line of sight, holding ε constant, while the *right column* shows variation around a sphere centered on the Sun, holding r (and therefore the illumination) constant. The *top* two plots show S/N variation for a hypothetical differential volume with unit electron density and the *bottom* two plots show S/N variation for a hypothetical self-similarly expanding volume (as in some CMEs)

Therefore units of elongation measured in coronagraphs are easily converted to units of distance via simply r (AU) $= \varepsilon$ (radians).

At larger angles from the Sun things become more complicated. Firstly, beyond around 15° from the Sun the small angle approximation no longer applies, leaving only the Point-P approximation (see, for example, [29]). Beyond angles of around 30° Point-P becomes unreliable, as the direction of propagation of the feature becomes important. One technique that has been used to overcome this problem is the so-called Fixed-ϕ approximation [35, 60], which uses geometry to match elongation measurements to a 3-D trajectory but assumes the feature is very small.

Beyond around $45°$ from the Sun, the geometry of the CME begins to become significant. Consider the situation illustrated in Fig. 3.4 which shows the situation for two very basic CME structures at large angles from the Sun. The assumed measured point on the CME is Q and the actual measured point is Q'. T' represents the point on the sphere where the LOS is tangent and crosses the CME, i.e., the observed leading edge. The distance of the assumed measured point from the Sun is r and the actual measured distance of this point is $r + \delta r$. This means that the difference between the measured distance and true distance is δr. Using this geometry, it can be shown that

$$r + \delta r = \frac{r}{2}(1 + \operatorname{cosec}(\lambda + \varepsilon)), \tag{3.15}$$

or

$$r = \frac{2R_0 \sin \varepsilon \operatorname{cosec}(\lambda + \varepsilon)}{1 + \operatorname{cosec}(\lambda + \varepsilon)}, \tag{3.16}$$

where R_0 is the distance from the observer to the Sun. Figure 3.4b also includes the angle α which is the angle the tangent to the CME makes to the Sun-observer line. When $\alpha + \varepsilon = 180°$ the leading edge of the CME has reached the observer, so beyond this point calculations of δr are meaningless.

For the case of the simple shell (Fig. 3.4c, d) there are three situations to consider:

$$r = R_0 \sin \varepsilon \qquad \text{when the LOS is a tangent to the front,}$$
$$\text{i.e., } 90° - (\lambda + \delta\lambda) < \varepsilon < 90° - (\lambda - \delta\lambda);$$
$$r = R_0 \sin \varepsilon \operatorname{cosec}(\varepsilon + (\lambda - \delta\lambda)) \text{ when the LOS contacts at } E_1,$$
$$\text{i.e., } \varepsilon > 90° - (\lambda - \delta\lambda);$$
$$r = R_0 \sin \varepsilon \operatorname{cosec}(\varepsilon + (\lambda + \delta\lambda)) \text{ when the LOS contacts at } E_2,$$
$$\text{i.e., } \varepsilon < 90° - (\lambda + \delta\lambda).$$
$$\tag{3.17}$$

The $\delta\lambda$ parameter is the semi-vertical angle of the cone swept out by the shell and $r + \delta r$ is the same as in Eq. (3.15).

Using these equations for the bubble and shell, Howard and Tappin [28] showed that for elongations where $\varepsilon + \lambda$ is near $90°$, the error in distance is relatively small, but as $\varepsilon + \lambda$ approaches $0°$ or $180°$ the apparent distance quickly diverges.

Chapter 5 of Howard [26] discusses the 3-D geometry of CMEs in more detail.

3.1.4 Summary

The detectability of a CME or any feature in the corona or solar wind is dependent on the following properties:

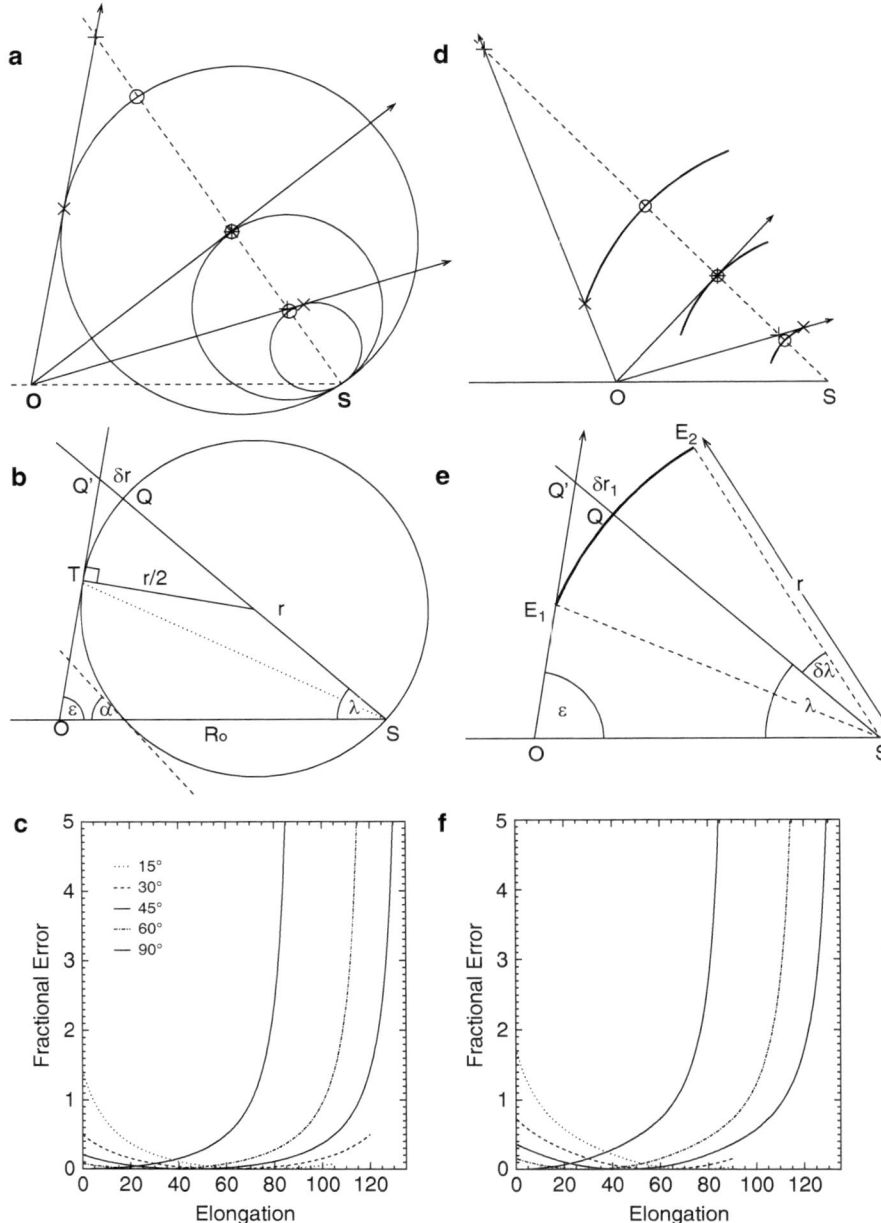

Fig. 3.4 Diagrams of two basic CME structures, (**a**)–(**b**) expanding bubble and (**d**)–(**e**) spherical shell with a semi-vertical angle of 30°. (**a**) and (**c**) The CME at three different locations during its expansion. The tangent drawn from the observer O across the CME surface shows the location of the relative leading edge. The × symbols represent the location of the leading edge seen by the observer, the ○ symbols show the true location of the leading edge at the central location, and the + symbols the inferred location of the leading edge based on the central location. (**d**) and (**e**). The

- The physics governing how the light is scattered off the particles in the feature (i.e., the physics of Thomson scattering);
- The observing properties of the detection instrument (sensitivity, field of view, etc.);
- The properties of the background noise (starfield, solar wind, zodiacal light, etc.);
- The geometrical structure and trajectory of the feature.

The characteristics of the solar wind and the nature of Thomson scattering produce a scattered feature intensity that is brightest when closest to the Sun, but the dropoff in intensity is very gradual until we are some considerable distance away from this point. The integrated line of sight across the entire solar wind introduces a noise contribution to the received signal as well, resulting in a peak in observed intensity that is closer to the observer than the closest point to the Sun. Finally, including a careful consideration of the geometry of the feature, we can extract information about the feature, including its size, shape, direction of propagation, speed, and mass.

3.2 Modeling CME Onset and Evolution

The first part of this chapter discusses the detection of CMEs and the means by which we may extract physical properties about them. It does not, however, consider the physics by which CMEs are launched from the solar atmosphere or how they evolve through the heliosphere. For the remainder of this chapter we discuss contemporary ideas behind the physics responsible for CMEs by way of the models that have been developed that attempt to describe this.

3.2.1 CME Onset

To date, we have been unable to observe the onset mechanism of CMEs directly. There are many phenomena that are associated with CME onset and early evolution (see, for example, Chap. 7 of Howard [26]), but it is unlikely that any of these are the onset mechanism of the CME itself. Some of them, however, are probably connected with it in some way, or caused by the same mechanism that launches the CME.

We know that the plasma β in the corona is low, meaning it is a region dominated by magnetic activity. So, it is reasonable to conclude that CME onset is a magnetic

Fig. 3.4 (continued) geometry allowing the derivation of the relationship between the difference in measured distance at a given point as a function of elongation [28]. (**c, f**) Plot of the fractional error $\delta r/r$ vs. elongation ε for a series of different directions λ. The larger the value the greater the error in determining the location of the CME by assuming the measurement is at the same location on the CME structure each time (Reproduced with kind permission of Springer Science and Business Media)

phenomenon. There are many models that describe the launch, formation and early acceleration of the CME but have mostly been unsuccessful at producing CME observations in coronagraphs. The advantage of using models is that boundary conditions can be adjusted until they match the observations. This is also their weakness, as often the conditions are adjusted without any real physical justification. Any ideal model of an onset mechanism must not only accurately describe the CME as it appears in coronagraphs, but also the related associated phenomena near and far from the Sun such as flares, filaments, interplanetary shocks and magnetic clouds. No model yet exists that is capable of achieving this objective and it is possible that there are different types of CME that are best described physically by different models.

For the purposes of onset and early acceleration description, we may divide CMEs into two types of eruption: gradual and explosive (although in reality these classes are not distinct). The first category involves CMEs that appear to take a long time to develop and move away from the Sun at a relatively gradual rate. The second type involves CMEs that have a high rate of acceleration and use large quantities of mechanical energy.

From Forbes et al. [19]:

> Most CME initiation models today are based on the premise that CMEs and flares derive their energy from the coronal magnetic field. The currents that build up in the corona as a result of flux emergence and surface flows slowly evolve to a state where a stable equilibrium is no longer possible. Once this happens, the field erupts. If the eruption is sufficiently strong and the overlying fields not too constraining, plasma is ejected into interplanetary space. If strong magnetic fields exist in the erupted region, then bright, flare-like emission occurs. (p. 254 [19])

The energy associated with coronal electric currents is called the "free" magnetic energy [38]. The CME's energy is predominantly provided by this free magnetic energy, and only the field associated with the corona is available to drive the CME. This energy builds up in the corona over time due to arising new magnetic fields and eventually, following a disruption (onset) of some kind, is released and the field erupts. This enables the emergence of new fields into a less energetic and complex region. Hence, the state of the coronal field before the launch of the CME governs its launch and behavior.

It is helpful to picture the pre-launch CME as two separate 3-D magnetic field structures, such as those shown in Fig. 3.5. In this view the structure that will become the CME is the blue-green (known as a core [53]) field which is held down by an overlying (red) field that straddles the core. When the CME launches it must either push the straddling field aside or stretch it out with it into the heliosphere.

The important factor is the angle between the magnetic field in the flux rope core and its axis. When the two are aligned, the tension force dominates, causing it to shrink and reach compact equilibrium. When they are at a significant angle, magnetic pressure destabilizes the flux rope and so expansion becomes energetically favorable. From that point, there are two fundamental questions to address when describing the physics of early CME evolution.

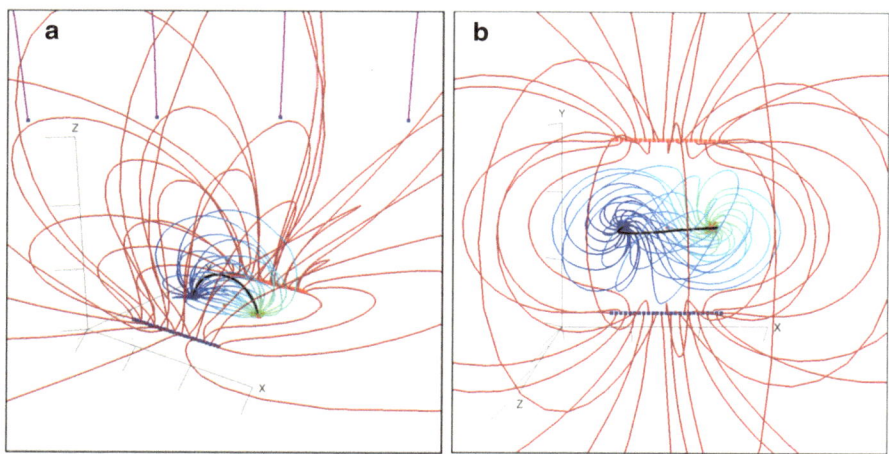

Fig. 3.5 A 3-D view of an ideal scenario involving a CME pre-launch magnetic field configuration. This consists of two structures: an underlying (core) field (*blue* and *green*) and an overlying straddling field (*red*) which acts to hold down the core. This is shown from two perspectives: (**a**) from an arbitrary 3-D direction, and (**b**) top-down, after a single twist has been introduced to the system (From Rachmeler et al. [57] and reproduced by permission of the AAS)

1. How can such large quantities of energy be delivered over such a short time frame?
2. Why would a CME expand if it was not energetically favorable to do so?

To the first question, consider only the mechanical energy involved. A fast CME travels in excess of 1,000 km/s and could have a mass of the order of 10^{13} kg. This speed is achieved within an hour of launch. That makes a power of no less than 10^{28} W during the early stages of CME launch, which must arise from the coronal magnetic field.

The second question has become known as the Aly-Sturrock energy limit, after Aly [1] and Sturrock [65] who theoretically demonstrated that the maximum energy state of any force-free magnetic field is the fully open one. This is an important revelation, as it states that if the surrounding magnetic field is stretched out with the CME then the resulting field would be more energetic than when it was in the pre-launch state. Hence, the CME would not spontaneously expand into this more energetic state. There are several means by which we may escape this paradox. Quoting Forbes et al. [19] again:

First, the magnetic field may not be simply connected and contain knotted field lines. Second, it may contain field lines that are completely disconnected from the surface. Third, an ideal-MHD eruption can still extend field lines as long as it does not open them all the way to infinity. Fourth, an ideal-MHD eruption may be possible if it only opens a portion of the closed field lines. Fifth, small deviations from a perfectly force-free state might make a difference. And finally, a non-ideal process, specifically magnetic reconnection, invalidates the constraint. (p. 255 [19])

Another alternative arises from more recent modeling in 3-D, which has demonstrated that the straddling field can also be pushed aside to make way for the erupting magnetic structure belief. This can occur with even the smallest disruption to the straddling field [57]. Which, if any, of these the Sun undergoes during the launch of the CME remains unknown. Nonetheless there are a number of theoretical means by which we may launch a CME without stretching the associated magnetic fields out to infinity.

3.2.1.1 Brief Review of Onset Models

Chapter 8 of Howard [26] provides a more detailed review of the models that we briefly summarize here. Good reviews CME modeling can be found in Forbes et al. [19] and Chen [9].

Contemporary models describing the onset and early evolution of CMEs can be divided into two categories: those requiring magnetic reconnection and those that do not. Magnetic reconnection (see Sect. 4.3) involves the merging of oppositely-directed magnetic fields and the release of energy. Those models not requiring magnetic reconnection include Magnetic Buoyancy, Flux Injection, Kink Instability and Mass Loading; those that require magnetic reconnection include Tether Cutting, Flux Cancelation, and Breakout.

Magnetic Buoyancy is based on the equilibrium between the expansion of heated coronal plasma into interplanetary space and the resistance of the coronal magnetic field against being opened [42]. The CME liftoff is a consequence of the opening of previously closed coronal magnetic fields and they are driven by the corona's natural tendency toward expansion. CMEs undergoing this mechanism tend to be slow [41, 43, 66, 78, 81]. Flux Injection (otherwise known as Toroidal Instability) treats the CME as a magnetic flux rope initially held in equilibrium, but which erupts as a result of poloidal magnetic flux being injected into the rope (see Fig. 3.6a). The flux rope is driven into the heliosphere via a driving Lorentz force between it and the surrounding field [7, 8]. Kink Instability involves the twisting (kinking) in the flux rope which increases in tension until a critical value is achieved, resulting in an instability [14, 15, 24, 69] (see Fig. 3.6b). The eruption is enabled by the formation of a vertical current sheet below the flux rope, following what is known as "supercritical twist". Finally, Mass Loading (Unloading) assumes the mass structures within filaments are sufficient to sustain the equilibrium between the gravitational binding energy and the magnetic forces expanding the prominence, enabling a build-up of mass and magnetic energy. When the mass structure is

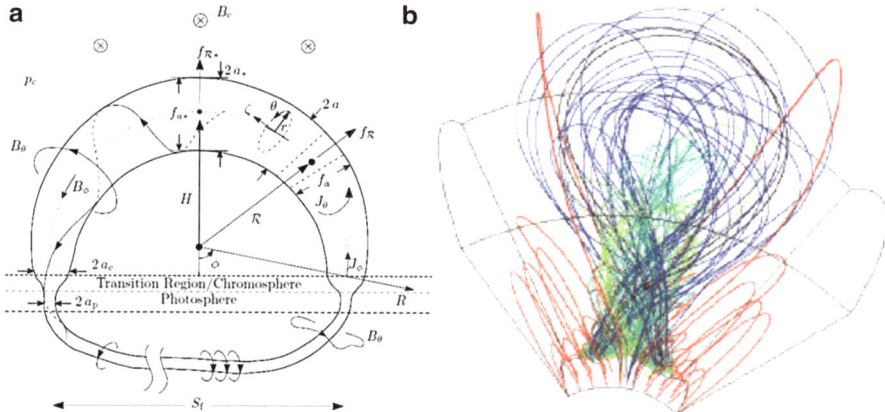

Fig. 3.6 (**a**) The conditions surrounding the flux injection model, modified from Chen [8]. The subscripts t and p refer to toroidal and poloidal respectively [58]. (**b**) A three-dimensional image of the coronal magnetic field via the Kink Instabilility Model [13]. The heavy *blue/green lines* represent the kinked flux rope, which erupts through the overlying magnetic structure (*red*), which is pushed aside (Reproduced by permission of the AAS)

disturbed such as via a reorientation of the magnetic field or a re-distribution of the mass, the equilibrium is broken and an eruption ensues [16, 36, 37, 39, 84]. All of these models describe a magnetic structure that is either held in equilibrium for a time and reorientated, and the energy release arises from the storage within the CME (and/or filament) magnetic structure.

The Tether Cutting (Tether Release) mechanism involves magnetic reconnection beneath the CME flux rope (the core field, which becomes sheared), thereby reducing the strapping tension force from the overlying coronal magnetic field [52, 53, 64] (see left panel of Fig. 3.7). The core field rises initially due to the imbalance between the pressure from the sheared field against the tension from the overlying field, and the surrounding field collapses into the vacancy beneath. Flux Cancelation (also known as Catastrophe) [48] involves the disappearance of magnetic fields at their separating neutral line. It begins with filaments, whose intrinsic fields can have a polarity opposite to that of the surrounding photospheric region. This enables field lines to close around the prominence material suspending it above the photosphere. Beneath this structure at the photosphere, magnetic flux is canceled by reconnection, a flux rope is formed [70], and a loss of equilibrium obtained [17, 18, 40]. Breakout [2] involves magnetic reconnection between the erupting core and the overlying strapping coronal magnetic field, enabling the delivery of extrinsic energy to the erupting core field (see right panel of Fig. 3.7). Originally, a large shear is introduced along the neutral line, opening the flux system which expands due to magnetic pressure from the photospheric shear. Eventually, a second reconnection process occurs at the base of the shear channel, and the original flux system is separated from the Sun by a current sheet. These mechanisms enable the delivery of large quantities of energy via the reconnection process.

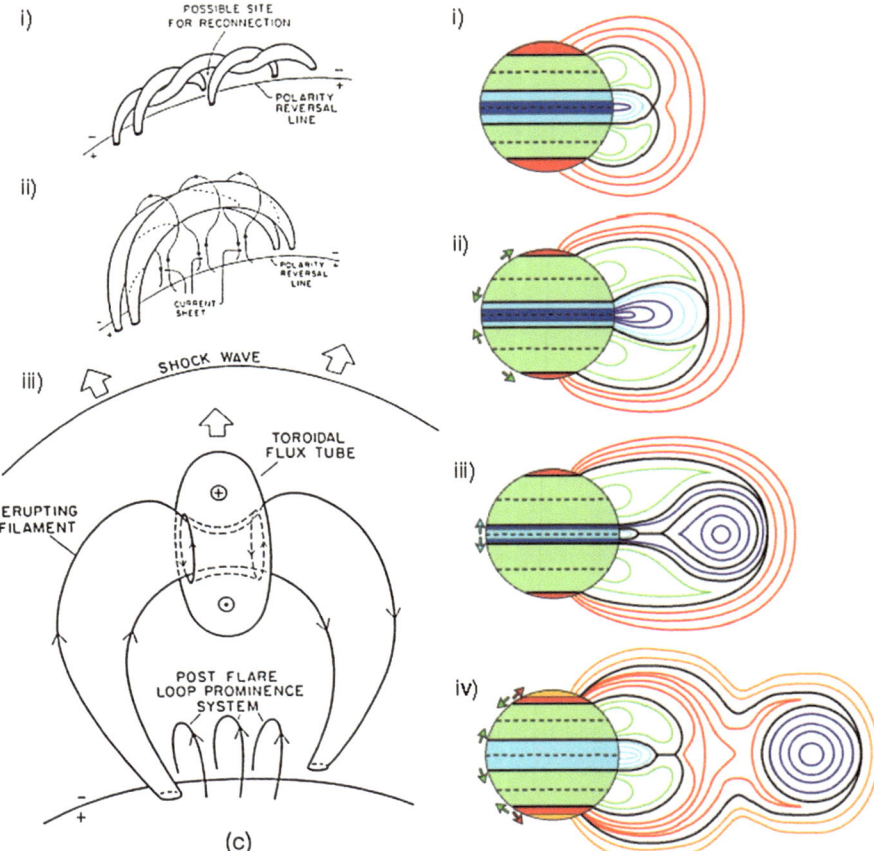

Fig. 3.7 (*Left*) Basic diagram of the Tether Cutting model according to Sturrock [64]. (**i**) The pre-launch magnetic field configuration associated with a prominence, showing where the reconnection may occur triggering the CME onset. (**ii**) The launch of the structure after magnetic reconnection and once the structure is no longer connected to the photosphere. (**iii**) Final form of the magnetic structure, following the onset. Reproduced with kind permission of Springer Science and Business Media. (*Right*) The Breakout model [46]. Diagrams of the main four stages of field evolution. (**i**) Initial topology, (**ii**) Shearing of the field and initiation, (**iii**) Flare reconnection starting deep in the shear channel and (**iv**) Reconnection allowing the relaxation of the field. The colours for the chosen fields indicate (*blue*) the central arcade straddling the equator, (*green*) two arcades associated with neutral lines at mid-latitudes, and (*red*) a polar flux system

All of the models summarized above provide energy to the CME via magnetic fields, some intrinsically, some extrinsically, and some (like Breakout) involve both. The Aly-Sturrock paradox can be overcome by moving the strapping overlying coronal field aside, and the vast quantities of energy required can be delivered by magnetic reconnection, or by the gradual storage of mass and energy. While it is generally accepted that an instability is probably involved in the onset process, no consensus has been reached on the fundamental mechanism required for that instability. Is magnetic reconnection required, or can the process be achieved without reconnection? If reconnection is involved, where and at what stage in the process does it occur? Can a single mechanism be used to describe all CMEs, or do different CMEs erupt via different mechanisms? And finally, how can we equate the associated phenomena with the launch mechanism? Given the nature of the Sun it is likely that more than one mechanism is responsible for CME onset and early acceleration. Further modeling and testing with observations continue.

3.2.2 CME Evolution Models

Once the CME is clear of the Sun the mechanism by which it propagates through the heliosphere is not well understood. This is for the same reasons as for the onset mechanism – sparsity of observational evidence and the flexibility of models. As with the launch, we do have physical constraints, physically reasonable assumptions, and some observations to provide us with clues. Much of the observational evidence has traditionally been from in-situ measurements and observations near the Sun such as coronagraphs. More recently we have had the advantages of heliospheric imaging, and there is also IPS to provide us with details of the CME density structure en route. To date however, as with the onset, CME evolution must be described using models with the advantages and disadvantages this brings.

Chapter 9 of Howard [26] reviews these evolution models and describes the solar wind through which the CMEs travel. We provide a brief review here. We divide the models into three categories:

1. **Aerodynamic Drag** assumes the behavior of the CME is entirely governed by its interaction with the solar wind and there is no significant intrinsic influence.
2. **Shock Dynamics** assumes the evolution of the CME is governed by the shock front formed ahead of the CME, as this is the only structure that interacts with the interplanetary medium in a significant way.
3. **Separate Ejecta** Treats the CME is an additional ejecta that is injected in the background solar wind. Some models regard the CME as a plasma ejecta, others treat it as a magnetic flux rope.

Each of these may be regarded as different aspects of essentially the same MHD problem, but with different bases and assumptions.

Firstly, let us review what information we have about CMEs in the solar wind. We know that beyond around 2 solar radii, the surrounding plasma β exceeds 1, meaning that the magnetic field plays less of a significant role than the plasma and surrounding fluid. This means that magnetic processes such as reconnection and magnetic pressure have less of an influence on the evolution of the CME than the hydrodynamics of the plasma. It is also known, however, that CMEs themselves are typically low-β phenomena (near 1 AU, values as low as 0.1 are common), so we have a typically low-β intrinsic magnetic field moving through a typically high-β environment. We know from coronagraph observations that shocks often form early in the evolution of the CME and that a well-defined shock with a build-up of compressed plasma is often observed ahead of the CME near 1 AU (Sect. 2.2.1). We also know that many CMEs at 1 AU contain highly structured magnetic flux ropes called magnetic clouds (Sect. 2.3.1). This means that the magnetic structure that is created near the Sun continues to maintain its structure at large distances. We also know about the ionic composition, temperature and bulk plasma speed of CMEs at 1 AU (Sect. 2.2.1), along with the behaviour of associated energetic particles, which provides empirical data on the CME there. Finally, with heliospheric imagers, interplanetary scintillation, and radio bursts, we can monitor the kinematic evolution and trajectory of CMEs (Sects. 1.5 and 3.1.3.3). Any working model that describes the evolution of CMEs through the interplanetary medium must not only appropriately explain all of the observations of CMEs, but also connect them with the Sun in a physically reasonable way, ideally connecting it with one of the launch mechanisms discussed in Sect. 3.2.1.

3.2.2.1 Aerodynamic Drag

Aerodynamic drag assumes the CME evolution is entirely governed by plasma and fluid dynamics (i.e., the magnetic field plays no role), and that CME dynamics always move to achieve kinematic equilibrium with the solar wind. The theoretical problem of evolution therefore becomes one of momentum transfer between the CME and its surroundings, and of kinematic equilibrium. This means that fast CMEs slow down and slow CMEs speed up until their speed matches that of the solar wind. Once there, they will simply cruise along at the solar wind speed, carrying accompanying structures along with them. Aerodynamic drag models typically regard the difference in speed between the CME and the solar wind as the dominating parameter. The tendency for fast CMEs to slow down and slow CMEs to speed up has been observed with coronagraphs [83] and interplanetary scintillation [47].

 Aerodynamic drag generally is divided into two scenarios: where solar wind material is allowed to accumulate with the CME and the other where it is not. The fundamental difference between the two is that the mass is a function of time with one model, where it remains constant with the other. The former is often referred to as the "Snow Plow" model as snow piling up in front of a snow plow is analogous to the solar wind piling up in front of the CME shock front. We refer to the latter simply as the "Drag" model. Both models consider the entire mass of the combined CME structure as observed in the heliosphere, that is they include material in the sheath region as well as the CME itself. Observational studies using these models include Cargill [6], Tappin [68], Howard et al. [30] and Vršnak et al. [74, 75].

3.2.2.2   Shock-Based CME Models

What the author calls shock-based CME models are those that regard the CME as a foreward shock wave moving through the solar wind. In such circumstances the physics of the CME are governed by those describing shock evolution, and so the CME is regarded as a perturbation in the surrounding medium. Further information on the theory of CME-propagated shocks can be found in Hundhausen [32]. While many of the models are originally based on the incorrect premise that the solar flare creates the CME, dynamics based on shock mechanics may not necessarily be an inaccurate description of CME evolution at large distances from the Sun. At the very least these models may be effective at describing the bright (dense) sheath structure that dominates what is observed by heliospheric imagers and interplanetary scintillation. Recall that the surrounding plasma β is high at these distances and so the dynamics are expected to be driven by the fluid, i.e., the mechanics of the shock may be appropriate for some CMEs. A number of these models have been developed over the decades [10,79], but we review here three of the more popular shock-based CME models.

The Shock Time Of Arrival (STOA) [11] and **The Interplanetary Shock Propagation Model (ISPM)** [63] drive a shock wave from a point source. They assume the flare is the source of the blast wave and use soft X-ray flare duration and Type II radio information as their boundary conditions. The background solar wind varies in the radial direction, but not in longitude, and the CME is introduced as a disturbance in an MHD solar wind. Results from these models include Dryer [11], Smart et al. [62], Smart and Shea [61], and Wu et al. [80]. The **Hakamada-Akasofu-Fry** kinematic solar wind model, version 2 **(HAFv2)** [20, 23] projects fluid parcels outward from the rotating Sun along fixed radials at successive time steps, in an inertial frame. If the speed gradient along a radial is steep enough, corotating interactions and interplanetary shocks are formed. This is how CMEs are worked into the background solar wind model – they are disturbances in the background medium driven by solar conditions. The initiation of the CME is achieved by modulating the inner boundary velocity field. Figure 2.9b shows results from the HAFv2 model for a CME observed in January 2007 [77]. Publications utilising HAFv2 include Fry et al. [20–22], McKenna-Lawlor et al. [49], Intriligator et al. [33], Howard et al. [30] and Webb et al. [76,77].

3.2.2.3   Separate Ejecta

Here we briefly discuss two models that treat the CME as a separate ejecta added to a background solar wind. The first introduces a dense structure with no intrinsic magnetic field. The second considers the CME as an expanding magnetic flux rope and considers the physics acting upon it from the surrounding medium, but does not produce a complete model of the solar wind. The first, the so-called ENLIL model is an important tool used to predict CME arrival times at the Earth by NOAA/SWPC. The latter is the extension of the Flux Injection model that we discussed in Sect. 3.2.1.1.

ENLIL [54,55] solves ideal MHD equations for plasma mass, momentum, energy density, and magnetic field. It is driven by the so-called WSA (Wang-Sheeley-Arge) empirical model [3] which uses ground-based magnetogram observations to feed a magnetostatic potential field source surface model, extending the coronal magnetic field out to 21.5 solar radii. ENLIL begins with a model of the solar wind, and the CME is introduced as an over-pressurized plasma cloud in addition to the background. CME parameters (e.g., location, size, speed) are estimated using models based on coronagraph observations of CMEs. The commonly-used version of ENLIL is therefore actually a combination of three models: WSA feeds the background solar wind parameters, which are evolved by ENLIL, while the CME cone model [82] introduces the CME. This combination is commonly used for space weather forecasting and is publicly available from the Community Coordinated Modeling Center (CCMC) web site (http://ccmc.gsfc.nasa.gov). Figure 2.9c shows ENLIL results for a CME observed in January 2007 [77]. Studies involving CMEs using ENLIL include Luhmann et al. [45], Odstrcil et al. [56], Taktakishvili et al. [67], Webb et al. [77] and Felkenberg et al. [12].

Chen [8] describes the extension of the Flux Injection model into interplanetary space. As described in Sect. 3.2.1.1 it begins with a toroidal flux rope in a surrounding poloidal field which is subjected to a Lorentz force. The equations become simplified as many of the forces at play near the Sun (such as gravity) disappear away from it, but these simplified equations are used to extend the CME into the heliosphere. It is through this model that a long-term driving force may act intrinsically within the CME, driving it out to large distances. This contradicts the view that the CME simply cruises into the solar wind once it is clear of the Sun.

3.3 Concluding Remarks

In order to obtain a complete understanding of CMEs, we must understand the physics that govern their onset, launch and evolution through the corona and heliosphere. We attempt to provide those descriptions using models based on our understanding of the corona and solar wind. Observations of CMEs inform those models, and so we must also understand the physics behind how CMEs are detected and how to interpret those images. This chapter is an abridged version of four chapters in the author's introductory book on CMEs (Howard [26]: Chaps. 4, 5, 8 and 9), and the reader is referred to those chapters, and references therein, for more information.

References

1. Aly, J.J.: Astrophys. J. **283**, 349–362 (1984)
2. Antiochos, S.K., DeVore, C.R., Klimchuk, J.A.: Astrophys. J. **510**, 485–493 (1999)
3. Arge, C.N., Pizzo, V.J.: J. Geophys. Res. **105**, 10465–10460 (2000)

4. Billings, D.E.: A Guide to the Solar Corona. Academic, New York (1966)
5. Brueckner, G.E., Howard, R.A., Koomen, M.J., Korendyke, C.M., Michels, D.J., Moses, J.D., Socker, D.G., Dere, K.P., Lamy, P.L., Llebaria, A., Bout, M.V., Schwenn, R., Simnett, G.M., Bedford, D.K., Eyles, C.J.: Solar Phys. **162**, 357–402 (1995)
6. Cargill, P.J.: Solar Phys. **221**, 135–149 (2004)
7. Chen, J.: Astrophys. J. **338**, 453–470 (1989)
8. Chen, J.: J. Geophys. Res. **101**, 27499–27520 (1996)
9. Chen, P.F.: Living Rev. Solar Phys. **8**, 1–92 (2011)
10. De Young, D.S., Hundhausen, A.J.: J. Geophys. Res. **76**, 2245–2253 (1971)
11. Dryer, M., Smart, D.F.: Adv. Space Res. **4**, 291–301 (1984)
12. Falkenberg, T.V., Vršnak, B., Taktakishvilli, A., Odstrcil, D., MacNiece, P., Hesse, M.: Space Weather **8**, S06004 (2010). doi:10.1029/2009SW000555
13. Fan, Y.: Astrophys. J. **630**, 543–551 (2005)
14. Fan, Y., Gibson, S.E.: Astrophys. J. **589**, L105–L108 (2003)
15. Fan, Y., Gibson, S.E.: Astrophys. J. **609**, 1123–1133 (2004)
16. Fong, B., Low, B.C., Fan, Y.: Astrophys. J. **571**, 987–998 (2002)
17. Forbes, T.G., Isenberg, P.A.: Astrophys. J. **373**, 294–307 (1991)
18. Forbes, T.G., Priest, E.R., Isenberg, P.A.: Solar Phys. **150**, 245–266 (1994)
19. Forbes, T.G., Linker, J.A., Chen, J., Cid, C., Kóta, J., Lee, M.A., Mann, G., Mikić, Z., Potgieter, M.S., Schmidt, J.M., Siscoe, G.L., Vainio, R., Antiochos, S.K., Riley, P.: Space Sci. Rev. **123**, 251–302 (2006)
20. Fry, C.D., Sun, W., Deehr, C.S., Dryer, M., Smith, Z., Akasofu, S.-I., Tokumaru, M., Kojima, M.: J. Geophys. Res. **106**, 20985–21002 (2001)
21. Fry, C.D., Dryer, M., Deehr, C.S., Sun, W., Akasofu, S.-I., Smith, Z.: J. Geophys. Res. **108**, 1070 (2003). doi:10.1029/2002JA009474
22. Fry, C.D., Detman, T.R., Dryer, M., Smith, Z., Sun, W., Deehr, C.S., Akasofu, S.-I., Wu, C.-C., McKenna-Lawlor, S.: J. Atmos. Solar-Terr. Phys. **69**, 109–115 (2007)
23. Hakamada, K., Akasofu, S.-I.: Space Sci. Rev. **31**, 3–70 (1982)
24. Hood, A.W., Priest, E.R.: Geophys. Astrophys. Fluid Dyn. **17**, 297–318 (1981)
25. Houminer, Z., Hewish, A.: Planet. Space Sci. **20**, 1703 (1972)
26. Howard, T.: Coronal Mass Ejections, An Introduction. Springer, New York (2011)
27. Howard, T.A., DeForest, C.E.: Astrophys. J. **752**, 130–142 (2012)
28. Howard, T.A., Tappin, S.J.: Space Sci. Rev. **147**, 31–54 (2009)
29. Howard T.A., Webb, D.F., Tappin, S.J., Mizuno, D.R., Johnston, J.C.: **111** (2006). doi:10.1029/2005JA011349
30. Howard, T.A, Fry, C.D., Johnston, J.C., Webb, D.F.: Astrophys. J. **667**, 610–625 (2007)
31. Howard, T.A., DeForest, C.E., Reinard, A.A.: Astrophys. J. **754**, 102–112 (2012)
32. Hundhausen, A.J.: Collisionless Shocks in the Heliosphere: A Tutorial Review, A87-25326 69–92, pp. 37–58. AGU, Washington, D.C.(1985)
33. Intriligator, D.S., Sun, W., Dryer, M., Fry, C.D., Deehr, C., Intriligator, J.: J. Geophys. Res. **110**, A09S10 (2005). doi:10.1029/2004JA010939
34. Jackson, J.D.: Classical Electrodynamics, 2nd edn. Wiley, New York (1975)
35. Kahler, S.W., Webb, D.F.: J. Geophys. Res. **112**, A09103 (2007). doi:10.1029/2007JA012358
36. Karpen, J.T., Antiochos, S.K., Hohensee, M., Klimchuk, J.A., MacNeice, P.J.: Astrophys. J. **553**, L85–L88 (2001)
37. Kippenhalm, R., Schlüter, R.: Z. Astrophys. **43**, 36 (1957)
38. Klimchuk, J.A., Sturrock, P.A.: Astrophys. J. **385**, 344–353 (1992)
39. Kuperus, M., Raadu, M.A.: Astron. Astrophys. **31**, 189 (1974)
40. Lin, J., Forbes, T.G., Isenberg, P.A., Demoulin, P.: Astrophys. J. **504**, 1006–1019 (1998)
41. Low, B.C.: Astrophys. J. **251**, 352–363 (1981)
42. Low, B.C.: Phys. Plasmas **1**, 1684–1690 (1994)
43. Low, B.C.: Solar Phys. **167**, 217–265 (1996)
44. Lugaz, N., Kintner, P., Möstl, C., Jian, L.K., Davis, C.J., Farrugia, C.J.: Solar Phys. **279**, 497–515 (2012)

45. Luhmann, J.G., Soloman, S.C., Linker, J.A., Lyon, J.G., Mikic, Z., Odstrcil, D., Wang, W., Wiltberger, M.: J. Atmos. Solar Terr. Phys. **66**, 1243–1256 (2004)
46. Lynch, B.J., Antiochos, S.K., DeVore, C.R., Luhmann, J.G., Zurbuchen, T.H.: Astrophys. J. **683**, 1192–1206 (2008)
47. Manoharan, P.K.: Solar Phys. **235**, 345–368 (2006)
48. Martin, S.F., Livi, S.H.B., Wang, J.: Aust. J. Phys. **38**, 929–959 (1985)
49. McKenna-Lawlor, S.M.P., Dryer, M., Smith, Z., Kecskemety, K., Fry, C.D., Sun, W., Deehr, C.S., Berdichevsky, D., Kudela, K., Zastenker, G.: Ann. Geophys. **20**, 917–935 (2002)
50. Mierla, M., Inhester, B., Antunes, A., Boursier, Y., Byrne, J.P., Colaninno, R., Davila, J., de Koning, C.A., Gallagher, P.T., Gissot, S., Howard, R.A., Howard, T.A., Kramar, M., Lamy, P., Liewer, P.C., Maloney, S., Marqué, C., McAteer, R.T.J., Moran, T., Rodriguez, L., Srivastava, N., St. Cyr, O.C., Stenborg, G., Temmer, M., Thernisien, A., Vourlidas, A., West, M.J., Wood, B.E., Zhukov, A.N.: Ann. Geophys. **28**, 203–215 (2010)
51. Minnaert, M.: Z. Astrophys. **1**, 209–235 (1930)
52. Moore, R.L., Roumeliotis, G.: In: Svestka, Z., Jackson, B.V., Machado, M.E. (eds.) Eruptive Solar Flares, pp. 69–78. Springer, New York (1992)
53. Moore, R.L., Hagyard, M.J., Davis, J.M.: Solar Phys. **113**, 347–352 (1987)
54. Odstrcil, D., Riley, P., Linker, J.A., Lionello, R., Mikic, Z., Pizzo, V.J.: In: Wilson, A. (ed.) Solar Variability as an Input to the Earth's Environment, p. 541. ESA SP-535, Estec, Noordwijk (2003)
55. Odstrcil, D., Pizzo, V.J., Linker, J.A., Riley, P., Lionello, R., Mikic, Z.: J. Atmos. Solar Terr. Phys. **66**, 1311–1320 (2004)
56. Odstrcil, D., Pizzo, V.J., Arge, C.N.: J. Geophys. Res. **110**, A02106 (2005). doi:10.1029/2004JA010745
57. Rachmeler, L.A., DeForest, C.E., Kankelborg, C.C.: Astrophys. J. **693**, 1431–1436 (2009)
58. Schuck, P.W.: Astrophys. J. **714**, 68–88 (2010)
59. Schuster, A.: Mon. Not. R. Astron. Soc. **40**, 35–57 (1879)
60. Sheeley, N.R., Jr., Walters, J.H., Wang, Y.-M., Howard, R.A.: J. Geophys. Res. **104**, 24739–24768 (1999)
61. Smart, D.F. Shea, M.A.: J. Geophys. Res. **90**, 183–190 (1985)
62. Smart, D.F., Shea, M.A., Barron, W.R., Dryer, M.: In: Shea, M.A., Smart, D.F., McKenna-Lawlor, S. (eds.) Proceedings of the STIP Workshop on Solar/Interplanetary Intervals, pp. 139–156. Book Crafters Inc., Chelsea, Maynooth, Ireland (1984)
63. Smith, Z., Dryer, M.: Solar Phys. **129**, 387–405 (1990)
64. Sturrock, P.A.: Solar Phys. **121**, 387–397 (1989)
65. Sturrock, P.A.: Astrophys. J. **380**, 655–659 (1991)
66. Sturrock, P.A., Weber, M., Wheatland, M.S., Wolfson, R.: Astrophys. J. **548**, 492–496 (2001)
67. Taktakishvili, A., Kuznetsova, M., MacNeice, P., Hesse, M., Rastätter, L., Pulkkinen, A.: Space Weather **7**, S03004 (2009). doi:10.1029/2008SW000448
68. Tappin, S.J.: Solar Phys. **233**, 233–248 (2006)
69. Titov, V.S., Démoulin, P.: Astron. Astrophys. **351**, 707–720 (1999)
70. van Ballegooijen, A.A., Martens, P.C.H.: Astrophys. J. **343**, 971–984 (1989)
71. van de Hulst, H.C.: Bull. Astron. Inst. Neth. **11**, 135–150 (1950)
72. van Houten, C.J.: Bull. Astron. Soc. Neth. **11**, 160–163 (1950)
73. Vourlidas, A., Howard, R.A.: Astrophys. J. **642**, 1216–1221 (2006)
74. Vršnak, B., Žic, T., Falkenberg, T.V., Möstl, C., Vennerstrom, S., Vrbanec, D.: Astron. Astrophys. **512**, id. A43 (2010)
75. Vršnak, B., Žic, T., Vrbanec, D., Temmer, M., Rollett, T., Möstl, C., Veronig, A., Čalogović, J., Dumbović, M., Lulić, S., Moon, Y.-J., Shanmugaraju, A.: Solar Phys. **285**, 295–315 (2013)
76. Webb, D.F., Howard, T.A., Fry, C.D., Kuchar, T.A., Mizuno, D.R., Johnston, J.C., Jackson, B.V.: Space Weather **7**, S05002 (2009). doi:10.1029/2008SW000409
77. Webb, D.F., Howard, T.A., Fry, C.D., Kuchar, T.A., Odstrcil, D., Jackson, B.V., Bisi, M.M., Harrison, R.A., Morrill, J.S., Howard, R.A., Johnston, J.C.: Solar Phys. **256**, 239–269 (2009)
78. Wolfson, R., Bongani, D.: Astrophys. J. **483**, 961–971 (1997)

79. Wu, S.T., Han, S.M., Dryer, M.: Planet. Space Sci. **27**, 255–264 (1979)
80. Wu, S.T., Dryer, M., Han, S.M.: Solar Phys. **84**, 395–418 (1983)
81. Wu, S.T., Guo, W.P., Wang, J.F.: Solar Phys. **157**, 325–248 (1995)
82. Xie, H., Ofman, L., Lawrence, G.: J. Geophys. Res. **109**, A03109 (2004). doi:10.1029/2003JA010226
83. Yashiro, S., Gopalswamy, N., Michalek, G., St. Cyr, O.C., Plukett, S.P., Rich, N.B., Howard, R.A.: J. Geophys. Res. **109** (2004). doi:10.1029/2003JA010282
84. Zhang, M., Low, B.C.: Astrophys. J. **600**, 1043–1051 (2004)

Chapter 4
Relevance to Space Weather

This chapter discusses the contribution of CMEs to space weather. We focus mostly on Earth, and briefly visit other bodies in the solar system such as planets and comets. The material presented here is an abridged and updated version of Chap. 11 from the author's introductory book on CMEs [34]. More in-depth reviews on the Earth's magnetosphere, those of the other planets and space weather can be found in Hargreaves [31], Kivelson and Russell [45], Song et al. [75] and Bothmer and Daglis [8]. An excellent, but slightly dated, online reference is the *Oulu Space Physics Textbook* which can be found at https://wiki.oulu.fi/display/SpaceWiki/Oulu+Space+Physics+Textbook.

When CMEs interact with bodies in the solar system a number of physical processes take place. The Earth's magnetosphere can be disrupted leading to what is known as a (geo)magnetic storm, but similar effects occur at other planets as well. The behavior of the magnetosphere and ionosphere are parts of the collective term known as space weather, but the most severe effects of space weather at the Earth are initiated by CMEs. It is the space weather impact of CMEs that workers are most interested in, and work continues to develop new and improved methods of determining the arrival time, speed and magnetic orientation of CMEs prior to their arrival at the Earth. This is an important part of space weather forecasting and its ultimate goal is to predict these parameters as early and accurately as possible. There are a number of space weather forecasting programs around the world, among them the Space Weather Prediction Center at NOAA in the US (http://www.swpc.noaa.gov/), and the Ionospheric Prediction Service in Australia (http://www.ips.gov.au/).

4.1 The Earth's Magnetosphere: Basics

The outer core of the Earth consists of molten metallic material, whose movement produces convection generating the geomagnetic field. If the Earth was in a vacuum the structure of this field would assume a form similar to that of a bar magnet (i.e., a dipole), and at lower altitudes and latitudes the geomagnetic field closely

T. Howard, *Space Weather and Coronal Mass Ejections*, SpringerBriefs in Astronomy, DOI 10.1007/978-1-4614-7975-8_4, © Timothy Howard 2014

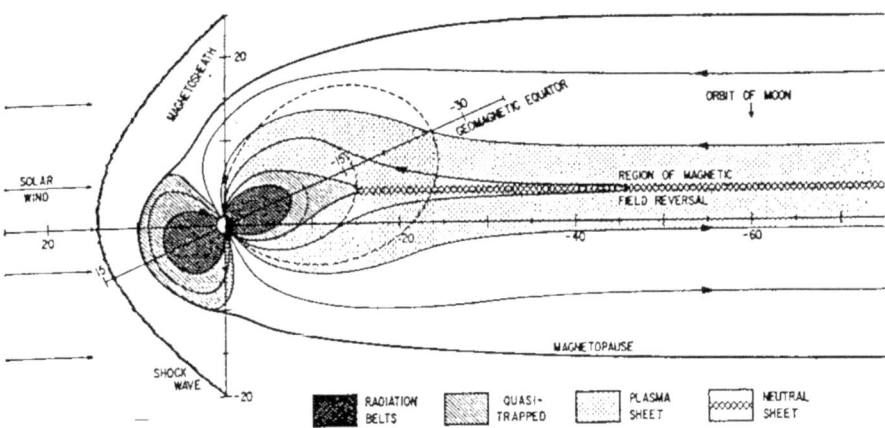

Fig. 4.1 An illustration of the geomagnetic field on the noon-midnight meridian [60] (Copyright Elsevier Academic Press, 1967)

resembles this. However, as with all bodies in the solar system the Earth lies in the solar wind which is a supersonic magnetized plasma. The resulting volume is called the magnetosphere and it extends out to millions of km into space [25, 65].

The solar wind exerts a pressure on the Sunward side (dayside) of the geomagnetic field, and because it is supersonic a bow shock forms ahead (Sunward) of the field. This causes a compression of the field there. Magnetic reconnection primarily on the dayside at high latitudes opens the field lines to the solar wind, and on the nightside the resulting volume extends out to at least 1,000 Earth radii [13].

Figure 4.1 shows an illustration of the geomagnetic field and some of its plasma regions. In those regions where there is no magnetic reconnection, the geomagnetic field cannot interact with the surrounding IMF, and so ions are trapped and confined to the magnetosphere. These field lines are called closed (the complete loop structures illustrated in Fig. 4.1). On the dayside, reconnection causes the merging of the geomagnetic field lines with the interplanetary field. These field lines are termed open, i.e., with only one end connected with the Earth (those lines that extend beyond the frame in Fig. 4.1). The regions at the base of these open field lines are called the polar caps, and a region called the cusp lies in between the last closed field line and the first open one. It is through the cusp region that the solar wind can be funneled into the magnetosphere to the ionosphere below. The aurora is caused by precipitating electrons from the solar wind through this region and along open field lines, energizing the atmosphere. A current is also created flowing westward in the Earth's equatorial plane at an altitude of 2–5 Earth radii. This is called the ring current, which is strongly enhanced during the main phase of a geomagnetic storm [11]. Table 4.1 outlines the important features of the magnetic components of the magnetosphere.

Table 4.1 Summary of the important features of the magnetosphere

Feature	Description
Bow shock	A shock formed when the solar wind encounters the geomagnetic field. It occurs on the dayside at a distance of around 15 Earth radii (R_E). Its thickness is of the order of 50–100 km
Foreshock	Region immediately Sunward of the bow shock populated by particles that leak from behind the bow shock. Particles arriving from the sunward direction are reflected off the bow shock and travel back toward the Sun. These are called back-streaming particles
Magnetosheath	Turbulent region immediately Earthward of the bow shock where the normal component of solar wind particle motions are decelerated to subsonic speeds
Magnetopause	Outer boundary defined on the dayside by the last closed magnetic field line and threaded elsewhere by field lines opened by reconnection.
Magnetotail	Region on the nightside of the magnetosphere, extending to over 1,000 R_E, containing both open and closed field lines
Cusp	Funnel-shaped region between the front and rear lobes of the magnetosphere. Through this region solar wind particles gain direct access to the ionosphere.
Auroral oval	Projection to the ground of the region where the auroral brightness maximizes
Polar cap	Projection to the ground of the region containing open field lines
Ring current	High energy ($\sim 10^{15}$ J) current located at around 2–5 R_E, flowing westward in the equatorial plane. The ring current is enhanced during a geomagnetic storm

Plasma enters the magnetosphere via a number of processes, namely diffusion from the solar wind, particles escaping from the ionosphere and magnetic reconnection. It is important to note that the footpoints of geomagnetic field lines are anchored to their position, so the field lines themselves corotate with the Earth. We can therefore represent the magnetosphere in terms of the plasma regions as well. They are shown in Fig. 4.2 and summarized in Table 4.2.

4.2 The Magnetospheres of Other Planets

As the other planets are also in the heliosphere, their magnetospheres are also governed by the solar wind and interplanetary magnetic field. The nature of the magnetic fields of the planets and their locations relative to the Sun are all different, so their responses to the solar wind and resulting magnetospheric configurations also differ greatly. Here we briefly review the nature of the magnetospheres of the other planets.

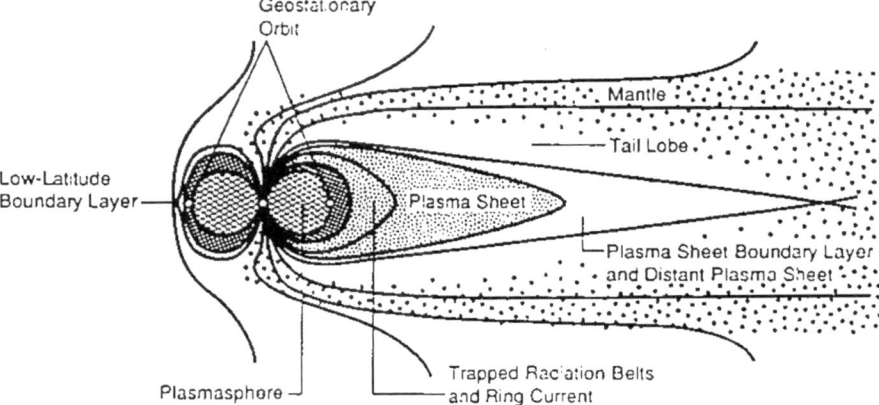

Fig. 4.2 Schematic diagram of plasma regions of the Earth's magnetosphere in the noon-midnight meridian [45]

Table 4.2 Plasma regions within the Earth's magnetosphere

Feature	Description
Plasmasphere	Cold, dense, corotating plasma forming a teardrop-shaped "sphere" around the Earth. Plasma accumulates in a region determined by a balance between the drift plasma, anti-Sunward solar wind pressure, plasma exchange with the ionosphere, and corotation of the plasma with the Earth
Plasmapause	Boundary between the solar wind driven convection in the outer magnetosphere and corotating plasma in the inner magnetosphere, typically at around 4 R_E
Plasma sheet	Hot accelerated plasma on closed field lines on the nightside
Tail lobes	Low density, cool plasma regions in the tail on open field lines on the nightside
Plasmatrough	Cold low-density plasma region just outside the plasmapause
Boundary layer	Region immediately Earthward of the magnetopause populated by a mixture of plasmas of solar wind and magnetospheric/ionospheric origin
Radiation belts	Two belts containing trapped particles. The inner belt consists mainly of high-energy protons produced when cosmic rays blast particles out of the upper atmosphere. The outer belt has high-energy electrons produced magnetospheric acceleration processes. The particles in these regions can easily move along magnetic field lines and so take the form of the surrounding field. Also called the Van Allen radiation belts

4.2.1   Mercury

Mercury has no ionosphere or atmosphere, but it does have a magnetic field which has a similar structure to that of the Earth's. It is a great deal weaker and therefore the magnetosphere is smaller than the Earth's relative to the size of the planet. Nevertheless, the iron core believed to produce this field is still large for a planet of Mercury's size. The field is a distorted dipole field like the Earth but it is met with solar wind of much larger pressure and field strength, and so the effects of space weather here are much more pronounced. NASA's *MESSENGER* spacecraft (launched August 2004) entered into orbit around Mercury in March 2011 and has been returning information on the Hermian magnetosphere since. A recent review of the Hermian magnetosphere can be found in Anderson et al. [2] and reports on the magnetosphere using *MESSENGER* include Slavin et al. [74], Boardsen et al. [7], Korth et al. [47] and Ho et al. [33].

4.2.2 Venus

Venus has no intrinsic magnetic field, presumably because it has no molten metallic core. Venus therefore does not deflect the solar wind and so its ionosphere is constantly stripped by the solar wind. Rather than a magnetopause, Venus has an ionopause, i.e., an interface between the ionosphere and the interplanetary magnetic field. The ionosphere of Venus therefore contains a great deal of plasma mantle, currents and induced magnetic fields, and is the source of large numbers of VLF and ELF waves. ESA's *Venus Express* (launched in November 2005) is currently in orbit around Venus. Publications regarding the Venetian "magnetosphere" include Bauer et al. [5], Phillips et al. [67] and Luhmann and Russell [53].

4.2.3 Mars

Mars has a very weak magnetic field which may be similar in generation to that of the Earth's moon. It does not seem to generate a magnetic field, and the weak field it does produce appears to arise from metallic bodies (e.g., rocks) which have been magnetized over time by their existence in an external magnetic field. This suggests that Mars may have once generated its own magnetic field. As with Venus, the solar wind often penetrates to the Martian atmosphere, but its effects are not as intense as on Venus, due to its larger distance from the Sun, where the solar wind pressure, density and the interplanetary magnetic field, are weak. Unlike with Venus, the Martian magnetic field does provide a small level of deflection of the solar wind. ESA's *Mars Express* (launched in June 2003) and NASA's *Odyssey*, *Reconnaissance Orbiter*, and *Mars Science Laboratory (MSL)* (launched April 2001, August 2005 and November 2011 respectively) are currently either in orbit around or on the surface of the planet. *MSL* is the most recent occupant of Mars, arriving in August 2012. *Phoenix* (launched October 2007) is no longer functional, being finally declared "dead" in May 2010. Further reading on the Martian magnetosphere may be sought from the *Space Science Reviews* volume on the subject [69].

4.2.4   Jupiter

Jupiter is not only the largest planet of the solar system, it also has the largest magnetic field. If the Jovian magnetosphere could be actually seen from the Earth, it would appear to be three times as large as the moon. It is considerably larger than a planet of its size and composition should be producing, meaning that there are additional characteristics of the Jovian environment that enhance its field. The Jovian core is in motion within a liquid metallic shell deep within the planet. It produces mostly a dipole field but has weaker quadrupole and octapole components, which act to enhance the field considerably [42]. A further enhancement is produced by one of its larger moons. Io is a highly active moon with constant eruptions of sulfur from its many volcanos and an ionosphere which is a fairly good electrical conductor. It orbits within the intense radiation belts of Jupiter, and the particles within collide with atoms in Io's atmosphere, spluttering the particles into a cloud of plasma around the planet [72]. The cloud becomes ionized and forms a torus around Jupiter. Both Jupiter and Io's torus are conductors, and so through the magnetic field lines there is a continuous current maintained between the planet and the moon. This changing current system produces an excellent dynamo between Jupiter and Io, and the Jovian magnetosphere is further enhanced.

Other interesting characteristics about the Jovian magnetosphere include:

- The field is oriented in the opposite direction to that of the Earth.
- The Jovian magnetopause is located at around 60 Jovian radii (R_J) and the bow shock is at around 80 R_J.
- The field is large enough to easily engulf the Sun if it could be placed within the field.
- Some of the Jovian moons (Io, Ganymede) may have magnetic fields of their own, carving mini-magnetospheres out of Jupiter's [36, 46, 71].

There are currently no spacecraft in orbit around Jupiter, since *Galileo* descended into the planet's atmosphere in September 2003. NASA/ESA's *Cassini* spacecraft made a flyby in late 2000-early 2001, as did NASA's *New Horizons* in February 2007. A new mission, called the Jupiter ICy moon Explorer (JuICE), and announced by ESA in May 2012, is planned for launch in 2022. Khurana et al. [42] provide a review of the Jovian magnetosphere and a book dedicated to the subject can be found in Dessler [17].

4.2.5   Saturn

The magnetic field of Saturn is produced in a similar fashion to that of the Jovian field, and although it is only a third as large it is still the second largest of the planets. Saturn provides a problem for some theorists because it has a magnetic axis almost exactly (within $1°$) aligned with the rotation axis. This is a problem because there is a well established theorem [14] stating that a planetary dynamo field can never be axially symmetric. Saturn's moon, Titan produces a torus around the planet similarly to Io around Jupiter. This is a hydrogen and nitrogen torus which acts to heat the plasma in Saturn's plasmasphere to an order of 10^6 K. The auroral ovals of Saturn have been observed in visible, infrared and ultraviolet light.

Cassini (launched in October 1997) is currently in orbit about the planet. *New Horizons* also made a flyby in June 2008. Reviews of the magnetosphere of Saturn include Russell and Luhmann [70], and more recently Mauk et al. [56] and Gombosi and Ingersoll [26].

4.2.6 Uranus

Uranus has a rotational axis that almost points at the Sun. It was therefore expected that the magnetic axis of Uranus should be close to its rotational axis and the Uranian cusp would be directly exposed to the Sun. This was found not to be the case. *Voyager 2* found that the magnetic axis was steeply inclined to the rotation axis (nearly 60°) causing it to spin like the axis of a top which is about to topple. The Uranian magnetosphere is therefore an extremely dynamic environment and it is expected that the plasma distribution in the surrounding field may not be as stable as the other planets. The field is believed to be generated by some mechanism located at relatively shallow regions in the Uranian atmosphere. Publications discussing the Uranian magnetosphere include Ness et al. [61] and Belcher et al. [6].

4.2.7 Neptune

Neptune's magnetic field is similar to that of Uranus in both generation mechanism and orientation. Its magnetic is inclined at $\sim$47° from the rotation axis. As the rotational axis of Neptune is around 30° from the ecliptic, this means that the magnetic axis is almost perpendicular to the Neptune-Sun plane. Triton, Neptune's moon, influences the behaviour of the outer magnetosphere [49]. As with Uranus, Neptune has only been investigated by *Voyager 2* and there are no known expeditions to this planet in the near future. Publications involving Neptune include Ness et al. [62] and Krimigas [48].

4.2.8 Pluto

It is unknown as to whether Pluto has a magnetosphere. While the *New Horizons* spacecraft (en-route to Pluto) does not have a magnetometer on board it does have an energetic particle detector (SWAP) that will provide information on the solar wind

Table 4.3 Comparison of the average locations of the bow shock and magnetopause, and the approximate angular difference between the rotational and magnetic axes (tilt)

	Bow shock	Magnetopause	Tilt angle
Earth	$15\,R_E$	$11\,R_E$	$10.8°$
Mercury	$1\,R_H$	$0.5\,R_H$	$10.0°$
Jupiter	$80\,R_J$	$45\,R_J$	$9.7°$
Saturn	$30\,R_S$	$21\,R_S$	$<1.0°$
Uranus	$30\,R_U$	$27\,R_U$	$59.0°$
Neptune	$30\,R_N$	$26\,R_N$	$47°$

around Pluto and its magnetosphere, if it has one. While we have no information as to the existence of the magnetosphere, some workers have attempted simulations of it. These include Delamere and Bagenal [16] and Harnett et al. [32].

Table 4.3 shows a comparison of the location of the bow shock and magnetopause relative to the radius of the planet, along with the difference between the rotational and magnetic axes.

4.3 Magnetic Reconnection

We have mentioned often in this book the importance of a process called magnetic reconnection with regard to CMEs. We have demonstrated how this process is important for their launch and early evolution (Sects. 1.7 and 3.2.1), and their interaction with the Earth and other magnetospheres (Sect. 4.1), and it has even been identified within CMEs en-route through the corona and heliosphere [27–29, 73]. Reconnection is therefore a vitally important physical process for space weather at the Earth and elsewhere. This section presents a basic review of magnetic reconnection in general and with regard to the Earth in particular. Excellent reviews of this process with regard to the Earth can be found in Kivelson and Russell [45] and Mozer and Pritchett [58].

The theory of magnetic reconnection as applying to solar flares dates back to the electromagnetic "neutral point" idea of Giovanelli [24], which was developed for MHD by Dungey and others through the 1950s [20, 21, 64, 76] and established by Petschek in 1964 [66].

When separate magnetized plasmas containing fields with antiparallel components move toward each other, the combined magnetic structure can be reconfigured to a lower energy state. Energy is therefore released during this process. The fields merge and the new configuration moves in a direction perpendicular to the original components. Figure 4.3 shows an illustration of this process.

Reconnection requires the nonphysical result in which the magnetic field points in two different directions at that same point in space. For this to occur there must be a non-zero electric field parallel to the magnetic field present that probably arises

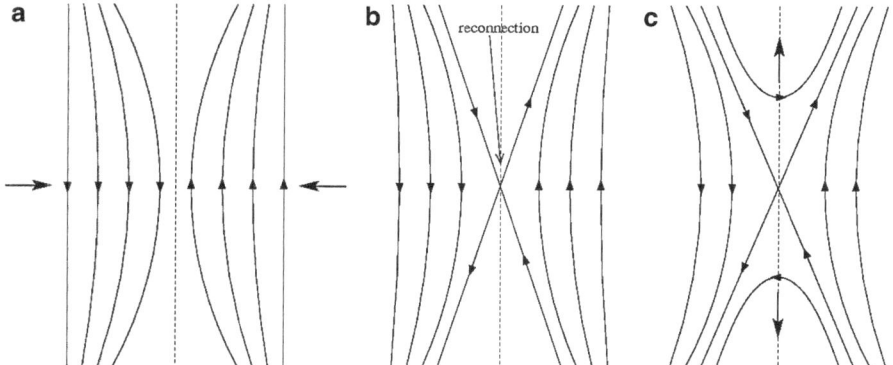

Fig. 4.3 2-D illustration of an ideal reconnection process, (**a**) before, (**b**) during, (**c**) after reconnection, where further reconnection processes can continue

from the parallel components of the electron pressure, inertia and/or resistivity. At the point where the field lines meet these properties allow the violation of the following equation

$$\mathbf{B} \times (\bigtriangledown \times \mathbf{E}_{||}) = 0, \tag{4.1}$$

where $\mathbf{E}_{||}$ is the parallel electric field. This enables reconnection to occur [58]. Magnetic reconnection can therefore be physically possible in a magnetized plasma if the above circumstances arise.

4.4 Magnetic Storms

In the previous section we established that magnetic reconnection can occur when two magnetized plasmas and their fields combine under certain conditions. These conditions can be met on the dayside of the Earth's magnetosphere. Recall that reconnection requires the fields in the separate plasmas to have anti-parallel components, so for reconnection to occur between a CME and the magneto-sphere, the north-south components of their fields must be oppositely directed. The geomagnetic field is from south to north, and so an CME with a strong southward field component will produce the strongest rate of reconnection with the magnetosphere [22]. Figure 4.4 shows a simplified diagram of how this may occur. Note that the green field lines (those connecting the Earth with the Sun and IMF) are in the polar cap region of the magnetosphere.

Magnetic reconnection between the CME and the geomagnetic field has two major effects on the magnetosphere. Firstly the reconnection process opens the geomagnetic field and connects it with the CME, allowing energetic particles

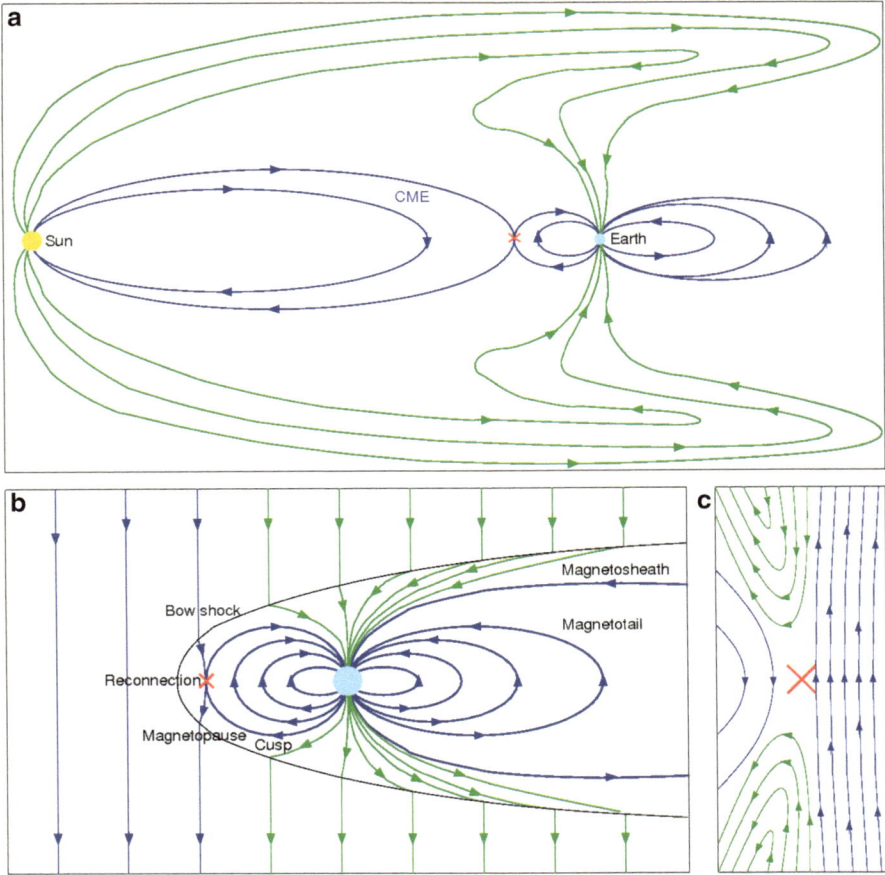

Fig. 4.4 Simplified depiction of magnetic reconnection on the dayside equatorial plane of the Earth's magnetosphere (modified from Mozer and Pritchett [58]). (**a**) Representation of the interplanetary magnetic field and CME that passes through the Earth and the Sun (*green*) and those that are connected only with the Sun or the Earth (*blue*). (**b**) A close-up view of the Earth's magnetosphere with the *green* and *blue lines* shown. Reconnection occurs at the point indicted where the northward-directed geomagnetic field meets the southward-directed interplanetary field. (**c**) Close-up view of the reconnection site on the dayside

contained within the CME to be injected into the magnetosphere. Secondly, the process releases energy, also injected into the magnetosphere. The result is a massive disruption to the magnetospheric system called a (geo)magnetic storm. Over 6 GW of power can be deposited into the ionosphere during an average storm [31] and for extreme cases, such as the Halloween event (October 2003), over 100 GW can be injected [18].

Recall also that along with the magnetic structure of the CME there is also the plasma component, which often produces a shock in the solar wind and exerts pressure on the magnetosphere when it encounters it. The ram pressure P is given by

$$P = \rho V^2 \tag{4.2}$$

where ρ and V are the density and speed of the CME. This pressure reduces magnetic field lines in size on the dayside, expanding the polar cap and causing the cusp to move further toward the equator. This results in the aurora being observed at even lower latitudes and an increase in the area of the ionosphere that is exposed to the solar wind. Also the shock and sheath contain a larger density and are moving faster than the ambient solar wind, which increases the concentration of energetic particles available to the magnetosphere. It is the arrival of an interplanetary shock that produces the so-called sudden (storm) commencement or S(S)C at the Earth.

4.4.1 The Effects of Magnetic Storms

In summary, a geomagnetic storm is the result of two effects: magnetic reconnection and increased ram pressure. The strength of the magnetic storm is therefore governed by two properties of the CME:

1. The magnitude and duration of the southward component of its magnetic field, at the point of contact with the magnetosphere;
2. Its ram pressure, which is a function of its speed and density.

This has four major effects at the Earth:

1. A large disturbance of the geomagnetic field, causing fluctuations in magnitude, direction and orientation;
2. An increase in the energetic particle population in the magnetosphere and ionosphere;
3. A reconfiguration of magnetic and plasma components of the magnetosphere, e.g., a shrinkage of the plasmasphere;
4. An intensification of the ring current and radiation belts,

which can result in the following:

- Electric charging of spacecraft circuitry, driving currents and short-circuiting electrical components on board [1, 3];
- Increased density and temperature in the ionosphere, increasing drag on spacecraft and advancing orbit decay [12];
- Electromagnetic induction in long electrical wires, leading to power station damage and failure [4];
- Electromagnetic interference leading to communications disruption [51, 77];
- Increased radiation dosage for high-latitude-flying aircraft passengers, staff and astronauts [23, 39].

Specific examples of assets that have most likely been damaged or destroyed by CME-induced magnetic storms include the *Galaxy 4*, *Galaxy 15* and *Equator-S* spacecraft (probably short-circuited by spacecraft charging), *Skylab* (brought down prematurely by increased atmospheric drag), the shutting down of the Hydro-Quebec power grid in Canada in 1989 and damage to a power grid in Sweden in 2003. Spacecraft launches have been delayed and aircraft routes changed as a result of magnetic storms. A review of the effects of magnetic storms can be found a report from a recent workshop on the topic [4].

4.5 CMEs at Other Bodies in the Heliosphere

Given the global nature of CMEs and their extent and influence on the Earth and solar wind, it seems likely that they encounter other bodies in the solar system, with similar major effects as well. Indeed, CMEs have been observed to impact other planets and possibly even comets in the solar system.

4.5.1 Planets

Because of its proximity to the Sun and weakness of its magnetosphere, Mercury suffers extreme effects of space weather. Its weak atmosphere is continually stripped away by the solar wind and photoionization processes. Models of the Hermian magnetospheric response to the solar wind have been produced [43, 44], but the author has been unable to identify a study involving a CME impacting the planet.

Venus does not have a magnetic field, and so CMEs act to enhance the already present atmospheric ionization [41], and the pressure pulse from the interplanetary shock decreases the size of the ionosphere [19], exposing even more of the neutral atmosphere to the solar wind [80]. CMEs have been observed at Venus by *Pioneer Venus Orbiter* [37, 52, 54, 59] and by *Venus Express* [55, 80].

Mars has only a very weak magnetic field, and so like Venus its ionosphere is constantly eroding in the solar wind. This effect is enhanced by the arrival of CMEs, but the effects are not as significant as at Venus. This is because Mars does have a magnetic field (albeit a weak one) and the CME itself is much weaker by the time it reaches Mars. It has been found for stronger CMEs that the magnetic field on the dayside becomes enhanced [15]. Studies of CMEs impacting Mars include McKenna-Lawlor et al. [57], Crider et al. [15] and Haider et al [30].

Once CMEs reach Jupiter and the outer planets they are in the region where they are beginning to interact with other CMEs and with other solar wind structures to form compressed regions called MIRs, and so it becomes increasingly difficult to identify individual events. Also, the strength of their magnetic fields are insignificant compared with those of the outer planets, and so their impact does not affect their behavior to any great extent. CME impacts have, however, been observed to have an impact on Jupiter and Saturn, primarily in the form of an aurora intensity enhancement [68].

4.5.2 Comets

CMEs may also be responsible for some types of comet disconnection events. Disconnection events occur when the tail of the comet appears to be disconnected from its head and moves independently through the solar wind [10, 40, 78]. Comet tails are now known to fluctuate in response to pressure changes in the local solar wind [50]. Three competing theories have been proposed to explain the triggering of cometry disconnection events [78]:

1. A sudden change in the ion production rate [79];
2. A sudden change in the solar wind pressure [35, 38];
3. Magnetic reconnection as the comet crosses the IMF sector boundary [9, 63].

Evidence from SMEI has suggested that in at least one case an CME may be responsible for a disconnection event. Kuchar et al. [50] identified six disconnections in comets NEAT and LINEAR and for one of them a faint CME was observed passing the comet tail at around the time of the disconnection. While not entirely conclusive, the evidence strongly suggests that CMEs interact with comet tails in the heliosphere as well at the Earth and other planets.

References

1. Allen, J.: Space Weather **8**, S06008 (2010). doi:10.1029/2010SW000588 (2010)
2. Anderson, B.J., Acuña, M., Korth, H., Slavin, J.A., Uno, H., Johnston, C.L., Purucker, M.E., Soloman, S.C., Raines, J.M., Zurbuchen, T.H., Gloeckler, G., McNutt, R.L., Jr.: Space Sci. Rev. **152**, 307–339 (2010)
3. Baker, D.N., Kanekal, S., Blake, J.B., Klecker, B., Rostoker, G.: EOS Trans. AGU **75**, 401 (1994)
4. Baker, D.N., Balstad, R., Bodeau, J.M., Cameron, E., Fennel, J.F., Fisher, G.M., Forbes, K.F., Kintner, P.M., Leffler, L.G., Lewis, W.S., Reagan, J.B., Small, A.A., III, Stansell, T.A., Strachan, L., Jr.: Committee on the Social and Economic Impacts of Severe Space Weather Events, NRC Workshop Report. Washington, DC (2009)
5. Bauer, S.J., Brace, L.H., Hunten, D.M., Intriligator, D.S., Knudsen, W.C., Nagy, A.F., Russell, C.T., Scarf, F.L., Wolfe, J.H.: Space Sci. Rev. **20**, 413–430 (1977)
6. Belcher, J.W., McNutt, R.L., Jr., Richardson, J.D., Selesnick, R.S., Sittler, E.C., Jr., Bagenal, F.: Uranus, pp. 780–830. University of Arizona Press, Tucson (1991)
7. Boardsen, S.A., Sundberg, T., Slavin, J.A., Korth, H., Soloman, S.C., Blomberg, L.G.: Geophys. Res. Lett. **37**, L12101 (2010). doi:10.1029/2010GL043606
8. Bothmer, V., Daglis, I.A.: Space Weather: Physics and Effects. Springer, New York (2007)
9. Brandt, J.C., Snow, M.: Icarus **148**, 52–64 (2000)
10. Brandt, J.C., Caputo, F.M., Hoeksema, J.T., Niedner, M.B. Jr., Yi, Y., Snow, M.: Icarus **137**, 69–83 (1999)
11. Burton, R.K., McPherron, R.L., Russell, C.T.: J. Geophys. Res. **80**, 4204–4214 (1975)
12. Compton, W.D., Benson, C.D.: Living and Working in Space: A History of Skylab. NASA SP-4208 (1983)
13. Cowley, S.W.H.: Planet. Space Sci. **39**, 1039–1047 (1991)
14. Cowling, T.G.: Mon. Not. R. Astron. Soc. **94**, 768–782 (1934)

15. Crider, D.H., Espley, J., Brain, D.A., Mitchell, D.L., Connerney, J.E.P., Acuña, M.H.: J. Geophys. Res. **110**, A09S21 (2005). doi:10.1029/2004JA10881
16. Delamere, P.A., Bagenal, F.: Geophys. Res. Lett. **31**, L04807 (2004). doi:10.1029/2003GL018122
17. Dessler, A.J. (ed.): Physics of the Jovian Magnetosphere. Cambridge University Press, New York (1983)
18. Dobbin, A.L., Griffin, E.M., Aylward, A.D., Millward, G.H.: Ann. Geophys. **24**, 2403–2412 (2006)
19. Dryer, M., Perez-de-Tejada, H., Taylor, H.A., Jr., Intriligator, D.S., Mihalov, J.D., Rompolt, B.: J. Geophys. Res. **87**, 9035–9044 (1982)
20. Dungey, J.W.: Philos. Mag. **44**, 725 (1953)
21. Dungey, J.W.: In: Lehnert, B. (ed.) Proceeding of the IAU Symposium, Paris, France, vol. 6, p. 135 (1958)
22. Dungey, J.W.: In: De Witt, C., Hieblot, J., Lebeau, A. (eds.) Geophysics: The Earth's Environment. Gordon Breach, New York (1963)
23. Dyer, C.: In: Sawaya-Lacoste, H. (ed.) Proceedings (1947) of the 2nd Solar Cycle Space Weather Euroconference, pp. 505–512. ESA, Noordwijk (2002)
24. Giovanelli, R.G.: Mon. Not. R. Astron. Soc. **107**, 338
25. Gold, T.: J. Geophys. Res. **64**, 1219–1224 (1959)
26. Gombosi, T.A., Ingersoll, A.P.: Science **327**, 1476–1479 (2010)
27. Gosling, J.T.: Proceedings of the Solar Wind 11 – SOHO 16, Whistler (2005)
28. Gosling, J.T., Skoug, R.M., McComas, D.J., Smith, C.W.: J. Geophys. Res. **110**, A01107 (2005). doi:10.1029/20043A010809
29. Gosling, J.T., Skoug, R.M., McComas, D.J., Smith, C.W.: Geophys. Res. Lett. **32**, L05105 (2005). doi:10.1029/2005GL022406
30. Haider, S.A., Abdu, M.A., Batista, I.S., Sobral, J.H., Kallio, E., Maguire, W.C., Verigin, M.I.: Geophys. Res. Lett. **36**, L13104 (2009). doi:10.1029/2009GL038694
31. Hargreaves, J.K.: The solar-terrestrial environment. Cambridge Atmospheric and Space Science Series, vol. 5. Cambridge University Press, Cambridge (1992)
32. Harnett, E.M., Wingless, R.M., Delamere, P.A.: Geophys. Res. Lett. **32**, L19104 (2005). doi:10.1029/2005GL023178
33. Ho, G.C., Krimigis, S.M., Gold, R.E., Baker, D.N., Slavin, J.A., McNutt, R.L., Jr., Winslow, R.M., Soloman, S.C.: J. Geophys. Res. **117**, A00M04 (2012). doi:10.1029/2012JA017983
34. Howard, T.: Coronal Mass Ejections, An Introduction. Springer, New York (2011)
35. Ip, W.-H., Mendis, D.A.: Astrophys. J. **223**, 671–675 (1978)
36. Isbell, D., Murrill, M.B.: Press Release, NASA, Galileo Finds Giant Iron Core in Jupiter's Moon Io, Webpage available via Views of the Solar System. http://www.solarviews.com/eng/galpr4.htm (1996). Cited 3 May 1996
37. Jian, L., Russell, C.T., Luhmann, J.G.: Solar Phys. **239**, 337–392 (2006)
38. Jockers, K.: Astron. Astrophys. Suppl. **62**, 791–838 (1985)
39. Jones, J.B.L.: In: Daglis, I.A. (ed.) Effects of Space Weather on Technology Infrastructure, pp. 215–234. Kluwer, Dordrecht (2004)
40. Jones, G.H., Brandt, J.C.: Geophys. Res. Lett. **31**, L20805 (2004). doi:10.1029/2004GL021166
41. Kar, J., Mahajan, K.K., Srilakshmi, M., Kohli, R.: J. Geophys. Res. **91**, 8986–8992 (1986)
42. Khurana, K.K., Kivelson, M.G., Vasyliunas, V.M., V.M., Krupp, N., Woch, J., Lagg, A., Mauk, B., Kurth, W.S.: In: Bagenal, F., Dowling, T.E., McKinnon, W.B. (eds.) Jupiter: The Planet, Satellites and Magnetosphere. Cambridge University Press, Cambridge (2004)
43. Killen, R.M., Potter, A.E., Reiff, P., Sarantos, M,. Jackson, B.V., Hick, P., Giles, B.: J. Geophys. Res. **46**, 20509–20525 (2001)
44. Killen, R.M., Sarantos, M., Reiff, P.: Adv. Space Res. **33**, 1899–1904 (2004)
45. Kivelson, M.G., Russell, C.T. (eds.) Introduction to Space Physics. Cambridge University Press, Cambridge (1995)
46. Kivelson, M.G., Warnecke, J., Bennett, L., Joy, S., Khurana, K.K., Linker, J.A., Russell, C.T., Walker, R.J., Polanskey, C.: J. Geophys. Res. **43**, 19963–19972 (1998)

47. Korth, H., Anderson, B.J., Raines, J.M., Slavin, J.A., Zurbuchen, T.H., Johnson, C.L., Purucker, M.E., Winslow, R.M., Soloman, S.C., McNutt, R.L., Jr.: Geophys. Res. Lett. **38**, L22201 (2011). doi:10.1029/2011GL049451

48. Krimigis, S.M.: Planet. Rep. **12**, 10–13 (1992)

49. Krimigis, S.M., Armstrong, T.P., Axford, W.I., Bostrom, C.O., Cheng, A.F., Gloeckler, G., Hamilton, D.C., Keath, E.P., Lanzerotti, L.J., Mauk, B.H., Van Allen, J.A.: Science **246**, 1483–1489 (1989)

50. Kuchar, T.A., Buffington, A., Arge, C.N., Hick, P.P., Howard, T.A., Jackson, B.V., Johnston, J.C., Mizuno, D.R., Tappin, S.J., Webb, D.F.: J. Geophys. Res. **113**, A04101 (2008). doi:10.1029/2007JA012603

51. Lanzerotti, L.J.: Space Storms Space Weather Hazards. In: Daglis, I.A. (ed.) Proceedings of the NATO Advanced Study Institute, Hersonissos, pp. 313–341 (2001)

52. Lindsay, G.M., Russell, C.T., Luhmann, J.G., Gazis, P.: J. Geophys. Res. **99**, 11–17 (1994)

53. Luhmann, J.G., Russell, C.T.: In: Shirley, J.H., Fainbridge, R.W. (eds.) Encyclopedia of Planetary Sciences, pp. 905–907. Chapman & Hall, New York (1997)

54. Luhmann, J.G., Kasprzak, W.T., Russell, C.T.: J. Geophys. Res. **112**, E04S10 (2007). doi:10.1029/2006JE002820

55. Luhmann, J.G., Fedorov, A., Barabash, S., Carlsson, E., Futaana, Y., Zhang, T.L., Russell, C.T., Lyon, J.G., Ledvina, S.A., Brain, D.A.: J. Geophys. Res. **113**, E00B04 (2008). doi:10.1029/2008JE003092

56. Mauk, B.H., Hamilton, D.C., Hill, T.W., Hospodarsky, G.B., Johnson, R.E., Paranicas, C., Roussos, E., Russell, C.T., Shemansky, D.E., Sittler, E.C., Jr., Thorne, R.M.: In: Dougherty, M.K., Esposito, L.W., Krimigis, S.M. (eds.): Saturn from Cassini-Huygens, pp. 281–331. Springer, Dordrecht (2009)

57. McKenna-Lawlor, S.M.P., Dryer, M., Fry, C.D., Sun, W., Lario, D., Deehr, C.S., Sanahuja, B., Afonin, V.A., Verigin, M.I., Kotova, G.A.: J. Geophys. Res. **110**, A03102 (2005). doi:10.1029/2004JA010587

58. Mozer, F.S., Pritchett, P.L.: Phys. Today (June edn.) 34–39 (2010)

59. Mulligan, T., Russell, C.T., Luhmann, J.G.: Geophys. Res. Lett. **25**, 2959–2962 (1998)

60. Ness, N.F.: In: King, J.W., Newmann, W.S. (eds.) Solar Terrestrial Physics. Academic, New York (1967)

61. Ness, N.F., Acuña, M.H., Behannon, K.W., Burlaga, L.F., Connerney, J.E.P., Lepping, R.P.: Science **233**, 85–89 (1986)

62. Ness, N.F., Acuña, M.H., Burlaga, L.F., Connerney, J.E.P., Lepping, R.P., Neubauer, F.M.: Science **246**, 1473–1478 (1989)

63. Neidner, M.B. Jr., Brandt, J.C.: Astrophys. J. **223**, 655–670 (1978)

64. Parker, E.N.: J. Geophys. Res. **62**, 509–520 (1957)

65. Parker, E.N.: J. Geophys. Res. **64**, 1675–1681 (1959)

66. Petschek, H.E.: In: Hess, W.N. (ed.) Physics of Solar Flares. NASA SP-50 (1964)

67. Phillips, J.L., Luhmann, J.G., Russell, C.T.: Adv. Space Res. **5**, 173–176 (1985)

68. Prangé, R., Pallier, L., Hansen, K.C., Howard, R., Vourlidas, A., Courtin, R., Parkinson, C.: Nature **432**, 78–81 (2004)

69. Russell, C.T. (ed.): Space Sci. Rev. **126** (2006)

70. Russell, C.T., Luhmann, J.G.: In: Shirley, J.H., Fainbridge, R.W. (eds.) Encyclopedia of Planetary Sciences, pp. 718–719. Chapman & Hall, New York (1997)

71. Sarson, G.R., Jones, C.A., Zhang, K., Schubert, G.: Science, **276**, 1106–1108 (1997)

72. Schneider, N.M., Bagenal, F.: In: Lopes, R.M.C., Spencer, J.R. (eds.) Io After Galileo: A New View of Jupiter's Volcanic Moon, pp. 265–286. Springer-Praxis, Chichester (2007)

73. Simnett, G.M.: Astron. Astrophys. **416**, 759–764 (2004)

74. Slavin, J.A., Acuña, M.H., Anderson, B.J., Baker, D.N., Benna, M., Boardsen, S.A., Gloeckler, G., Gold, R.E., Ho, G.C., Korth, H., Krimigis, S.M., McNutt, R.L., Jr., Raines, J.M., Sarantos, M., Schriver, D., Solomon, S.C., Trávníček, P., Zurbuchen T.H.: Science **324**, 606–610 (2009)

75. Song, P., Singer, H.J., Siscoe, G.L.: Space Weather. AGU, Washington, D.C. (2001)

76. Sweet, P.A.: In: Lehnert, B. (ed.) Proceedings of the IAU Symposium, vol. 6, p. 123 (1958)
77. Thompson, R., McDonald, A.: Communications and Space Weather, Webpage available via IPS (Australia). http://www.ips.gov.au/Educational/1/3/4 (2007). Cited 2007.
78. Voelzke, M.R.: Earth Moon Planets **90**, 405–411 (2002)
79. Wurm, K., Mammano, A.: Astron. Astrophys. Suppl. **18**, 273–286 (1972)
80. Zhang, T.L., Pope, S., Balikhin, M., Russell, C.T., Jian, L.K., Volwerk, M., Delva, M., Baumjohann, W., Wang, C., Cao, J.B., Gedalin, M., Glassmeier, K.-H., Kudela, K.: J. Geophys. Res. **113**, E00B12 (2008). doi:10.1029/2008JE003128

Chapter 5
Recent Developments

In many regards, the view of coronal mass ejections today has remained largely as it has been since the resolution of the Solar Flare Myth debate in the early 1990s (see Sect. 2.4). Then as now, the CME was regarded as a large magnetic structure that is likely formed in the solar corona, which then erupts (sometimes explosively) into the solar wind, carrying large quantities of magnetic field and energy away with it. This Brief explores the CME as it has been observed historically (Chap. 2), how it is detected and how information is extracted from it (Chap. 3). We discuss many of the models that describe the onset and evolution of CMEs (Chap. 3) and their effects on space weather (Chap. 4), which at the Earth can be very dramatic.

Much of the first four chapters of this Brief are abridged and updated versions of the author's introductory book on CMEs [19], which discussed the status quo regarding our understanding of CMEs up until around 2010. Since that time, research work into CMEs and space weather has continued to progress and our picture of their onset and evolution continues to gradually be brought into sharper focus. In this chapter we discuss some of the developments that have emerged since the writing of the introductory book (i.e., since 2010). We focus on observational rather than modeling results, since although the models discussed in Chap. 3 continue to be refined and developed, to the author's knowledge no new models have emerged that alter the fundamentals described in those we have already discussed. Observationally, there has been nothing that could be described as a "breakthrough" either, but some issues that had been debated over the years have now been resolved. These developments have arisen primarily from the most recent solar-observing spacecraft missions: *STEREO*, and *SDO*.

5.1 Mission Updates Post-2010

Chapter 3 of Howard [19] presents a list of the spacecraft missions throughout the decades that have made significant contributions towards our understanding of CMEs. Those listed there as still operational were the *Voyagers*, *Wind*, *SOHO*,

T. Howard, *Space Weather and Coronal Mass Ejections*, SpringerBriefs in Astronomy,
DOI 10.1007/978-1-4614-7975-8_5, © Timothy Howard 2014

ACE, TRACE, RHESSI, Coriolis, Hinode, STEREO, and *SDO.* Since then, *TRACE* has concluded (June 2010) and SMEI, which was on board *Coriolis,* has also been shut down (September 2011). Two of the instruments on board *SOHO* have been turned off as they were regarded as obsolete compared with the new generation of instruments. These are EIT (July 2010) and MDI (April 2011). CDS was planned for shutdown in 2011 (but it hasn't happened yet), and SUMER is planned to be turned off following the successful launch of a new mission IRIS, scheduled for December 2012. No replacement spacecraft coronagraph has yet replaced LASCO and so it continues to function and despite its 17 years in space, operates well.

The twin *STEREO* spacecraft continue to be operational but in their seventh year are now well behind the Sun relative to the Earth (see Fig. 2.10). This limits their ability to observe Earth-directed CMEs but they have provided complete coverage of the entire solar globe for the first time. This was accomplished in mid-2011 and continues today: the regions on the Earth side of the Sun that are now blind to *STEREO* are observed by *SDO.*

There have been no new solar or CME observing spacecraft missions since *SDO* was launched in February 2010, but the sounding rocket missions *SUMI* (July 2010) [33], *RAISE* (August 2010) [17, 18], and *Hi-C* (July 2012) [12] have briefly provided high quality solar images in the interim, and IRIS is planned for launch later in 2012. No new coronagraph missions are planned until Solar Orbiter [36] and Solar Probe Plus [39], currently scheduled for launch in 2017 and 2018 respectively, by which time LASCO, if it remains functioning, will have been operational for 22 years.

5.2 Latest Observational Work

Data from missions past and present are archived and work continues with those datasets today (see Sect. 2.8 for some examples). Here we focus on results related to CMEs that have arisen from the two most recent CME-observing spacecraft missions: *STEREO* and *SDO.*

5.2.1 STEREO *Observations*

The main component of the *STEREO* mission design was to enable 3-D reconstruction work from the multiple viewpoints enjoyed by combined measurements (both in-situ and imaging) from both the *STEREO* spacecraft and the observatories and probes near the Earth. Not surprisingly, a vast amount of work has been done on 3-D reconstruction of CMEs, mostly using triangulation or basic geometrical simulation from coronagraph images. The 3-D geometry work began with Mierla et al. [30] for the COR1 coronagraphs and Howard and Tappin [21] for COR2, and the number of 3-D reconstruction tools has been rapidly growing since. The review paper by Mierla et al. [31] discussed many of these reconstruction techniques as they were in 2010, and since then they have become more sophisticated [2, 6, 10, 32, 46]. Likewise, the 3-D reconstruction techniques using heliospheric imager data have also improved [1, 5, 26, 34, 43].

The problem with 3-D reconstruction work lies partly in the optically thin nature of CMEs. Geometric triangulation relies entirely on the measured point from different viewpoints lying exactly in the same place in 3-D space. If this assumption breaks down, 3-D triangulation becomes meaningless. We already know (see Sect. 3.1.3.3) that at large angles from the Sun the structure of the CME changes the location of the line of sight crossing with the CME at different times, but also from different locations; so it is nonsense to attempt 3-D triangulation using heliospheric imager data. It is also a questionable technique using coronagraphs for geometrical reasons, but because of different effects. Here the optically thin nature of CMEs creates a tendency for an observer to place the measured location of the CME structure toward the far end of the line of sight. This means that the best we can do to locate a leading edge on a CME is to place it somewhere within a polygon bound by the four lines of sight passing from two separated observers across each edge of the CME (see Fig. 5.1a for the geometry). As a result, it is impossible to identify a common point in 3-D space when observed from two different viewpoints [22]. Exceptions would include very small CMEs or those with a geometry such that the crossover point of the lines of sight with the CME are close to the far edge. It would therefore seem that the most appropriate 3-D reconstruction techniques of CMEs from coronagraphs are those that place the CME within a polygon, as there simply is no more 3-D information about the CME within the coronagraph images. Reports on techniques that apply this assumption to coronagraph data include de Koning et al. [7] and Feng et al. [10]. Figure 5.1b shows one such reconstruction from de Koning et al. [7].

5.2.1.1 Scientific Developments

Sadly, very little actual science has emerged from the vast assortment of new analytical techniques discussed in Sect. 5.2.1. That is to say, while we can now analyze the heck out of CMEs, 3-D reconstruction efforts using *STEREO* have provided no new information about the science behind CMEs than we already knew before *STEREO* was launched. The good news is that finally, some real scientific results are now beginning to emerge.

The most notable new CME scientific results to emerge from *STEREO* are not from the 3-D reconstruction work, but rather from the capability of the SECCHI imager suite to continuously track CMEs, related features, and other heliospheric transients from the Sun through the corona and inner heliosphere. For the first time, for example, we can now easily identify the signatures of corotating interaction regions in the SECCHI dataset, which have often previously been confused with CMEs. The interaction of CMEs in the heliosphere is being observed and studied in

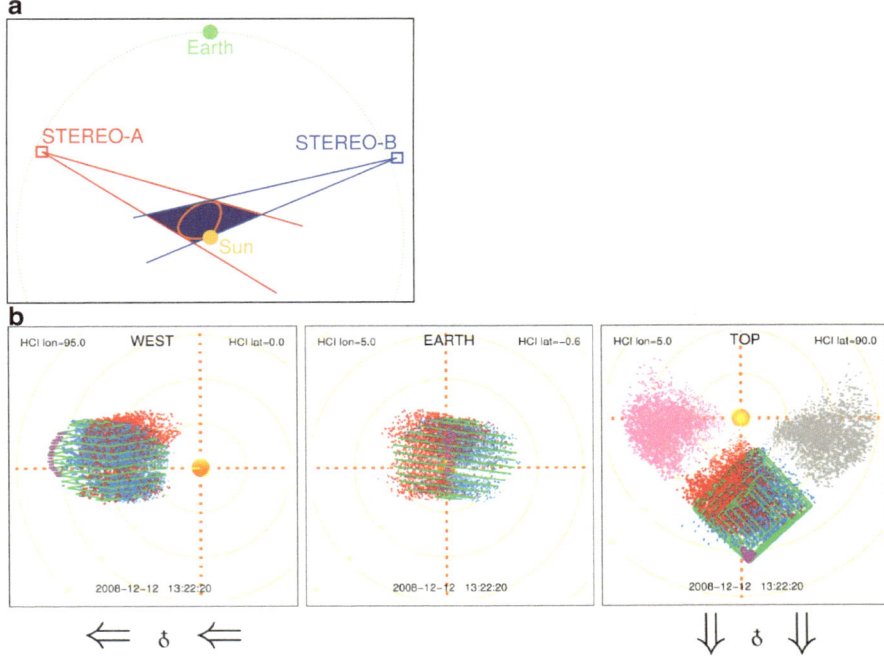

Fig. 5.1 (**a**) Diagram of a CME (*pink arc* extending from the Sun) observed by both *STEREO* spacecraft when they were around 45° from the Sun-Earth line [7]. Two lines of sight (*red* from *STEREO-A*, *blue* from *STEREO-B*) cross the CME at a tangent at both ends – these represent the apparent leading edge of the CME. Given the optically thin nature of the CME, the best we can do regarding 3-D reconstruction is to place the CME somewhere within the polygon bound by the four lines of sight (the *purple* region). (**b**) 3-D reconstruction of a CME observed in December 2008 [6] from three different viewpoints. In each panel, the direction to the Earth is indicated with the *arrows*. The *green polygons* represent the reconstruction using the polygon technique of de Koning et al. [7], while the dots are the results from a separate 3-D location technique using polarimetry (see de Koning and Pizzo [6]). The 3-D reconstruction techniques place the CME almost directed straight towards Earth

detail for the first time, and recent data processing developments have enabled the tracking of the anatomy of CMEs from the Sun through to the Earth.

Close to the Sun, the overlap between the fields of view of the solar disk imagers on board *STEREO* (EUVI) and the innermost coronagraph COR1 has enabled the study of CME onset. Recall that on *SOHO*, the innermost coronagraph on LASCO (C1) was disabled early in the mission, meaning that such studies could not be conducted. Patsourakos et al. [37] applied a simple model to an accelerating CME, tracking it from its origins in EUVI into the COR1 field of view. They found that the CME exhibited two phases of expansion: an initial overexpansion (compared to its rise); followed by a phase of nearly self-similar expansion. CME initial acceleration has also been reported by Temmer et al. [40], who measured the kinetics of the impulsive acceleration phase of the CME up to 4 solar radii, and found a correspondence between the acceleration profile of a number of CMEs and

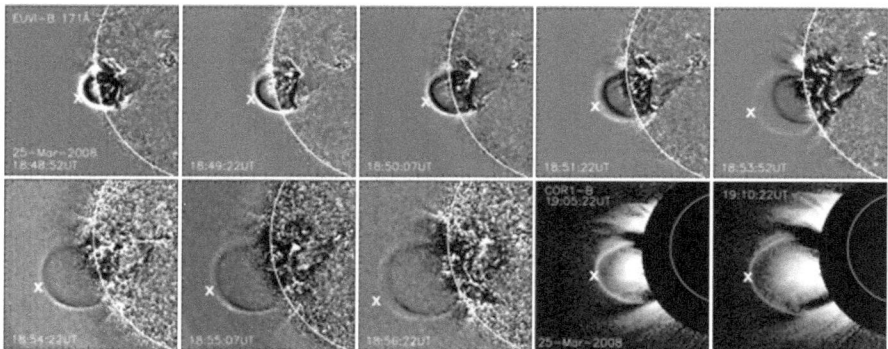

Fig. 5.2 Observations of the launch of a CME (from Fig. 6 of Temmer et al. [40]), first immediately off the solar disk observed by EUVI, then in the corona by COR1. The white "×" denotes the location at which kinematic measurements were made for the CME

the energy released by the associated solar flare. They conclude that these results support the so-called "standard" CME/flare model, which describes a feedback relationship between the CME acceleration and the flare energy release. Figure 5.2 shows the tracking of the launch of the CME from EUVI into COR1.

At large distances from the Sun, it has been believed for some time that some CMEs must somehow merge or overlap, given the proximity of many CMEs and the variability of their kinematics (e.g., fast CMEs following slow CMEs). The idea of interacting CMEs has been around for at least 10 years. From 2001, Gopalswamy and colleagues [13–15, 29] have associated radio bursts, interplanetary shocks, and solar energetic particle events with interacting CMEs, and they have also been modeled by a number of workers [25, 42, 44]. However it was not until *STEREO* that their merger could be directly observed. In a global effort to study a group of CMEs that were launched in August 2010, a number of workers discussed what appeared to be the interaction between two of a group of at least four CMEs. This collection of CMEs was observed by a multitude of spacecraft, offering the largest in-situ and imaging coverage yet seen. Studies on these events include Liu et al. [24], Temmer et al. [41], Harrison et al. [16] and Möstl et al. [35]. Figure 5.3 shows two diagrams of two different events. They found, along with a recent study by Lugaz et al. [27] involving interacting CMEs observed in May 2010, that a fast CME ran into a slower CME ahead of it. The fast CME decelerated upon interaction and that the two were deflected. While these interactions have been confirmed in observation, questions remain elusive regarding the physical description as to what actually happens when CMEs interact. What happens to their flux ropes, for example. Do they merge, or are they simply layered alongside each other [4]? If each drives a shock, do these shocks interact, enhance, or combine together? How about

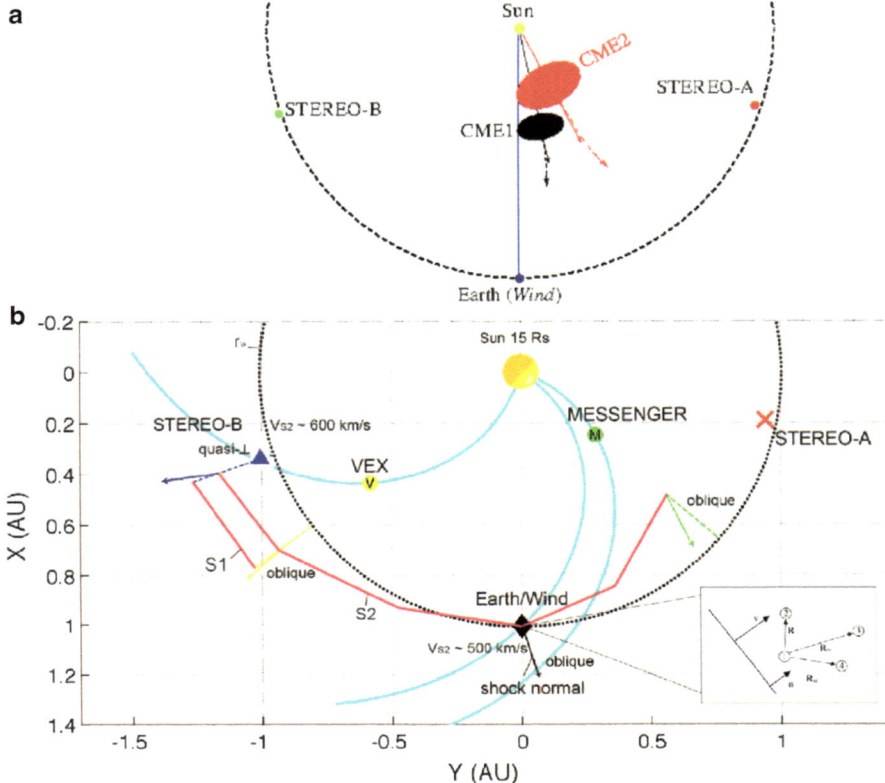

Fig. 5.3 Illustration of the locations and trajectories of two CMEs observed in May 2010, from (**a**) Fig. 5 of Lugaz et al. [27]. (**b**) Global configuration of two shock fronts S1 and S2 at their estimated locations at August 3, 2010 at 17:05UT (from Fig. 2 of Möstl et al. [35]). The approximate shape of the shocks (*red solid lines*) is shown by estimating the heliocentric distance of the shocks at each location for the given point in time. For S2, shock normals are drawn as *arrows* in the color consistent with the spacecraft position, the configuration of the shock is indicated as oblique or quasi-perpendicular. The nominal Parker spiral field is drawn connecting the Sun to each spacecraft for the upstream conditions of solar wind speed 450 km/s at *STEREO-B* and *VEX* and 400 km/s at *Wind* and *MESSENGER*. The insert illustrates the four-spacecraft method to determine the shock normal orientation near Earth

the magnetic connectivity with the Sun for each separate CME? These and other questions will doubtless be addressed in future studies regarding this (relatively) new field of CME research.

Recently, a new data processing pipeline was developed for *STEREO*/HI [8]. This reduced background starfield and zodiacal light considerably, enabling the clear observation and tracking of features at lower intensities than had been possible before. When applied to the rest of SECCHI [20], the ability to isolate and track components of CMEs through the inner heliosphere was made possible. The flux rope component that comprises the magnetic cloud observed in-situ (Sects. 1.5 and 2.3.1) often comprises of a region of reduced density, which creates an apparent void region in white light images. This void was tracked by Howard and DeForest [20] from the magnetic cloud signature through the heliosphere back to their coronagraph origins, confirming the long-suspected belief [11] that the cavity component of the three-part CME (Sect. 1.5) is the feature that becomes the magnetic cloud. Once confirmed, it was possible to investigate the structural evolution of the flux rope, and it was found to become highly distorted, strongly suggesting that the flux rope was not force-free. Figure 5.4 shows a CME observed with the SECCHI pipeline and an elongation-time "J-map" showing the evolution of a cross-section of the CME structure. Work continues making the connection between CME components and their in-situ signatures [22], including one study involving the presence of solar-connected strapping field for one CME at a distance of 1 AU from the Sun [9], which seems to validate the Aly-Sturrock energy limit (Sect. 3.2.1) for at least one type of (slow) CME. Figure 5.5 shows a diagram of the magnetic structure of the pre-eruptive CME state and the state at a large distance from the Sun, derived from these results.

5.2.2 SDO *Observations*

SDO does not actually observe the CME, or at best it can only observe it at its earliest stages, so observations using *SDO* have focused more on activity related to CMEs, such as solar flares and filaments. Liu et al. [23], for example, investigated CME-related EUV waves, while Zhang et al. [45] reported on the evolution of a magnetic flux rope leading to the launch of a solar flare. However, attempts to study the CME directly with *SDO* have been made. Early results include Patsourakos et al. [38], who analyzed the formation of what they called the CME EUV bubble and measured its initial dynamics and thermal evolution. They report that their observed CME formed through three phases: a slow self-similar expansion, followed by a fast but short-lived period of strong lateral overexpansion, and then a state of self-similar expansion until it left the *SDO* field of view. Figure 5.6 shows the formation of this CME as reported by Patsourakos et al. Other works include a study on the formation of a shock wave on both the solar disk and off the limb by Ma et al. [28], who found it to propagate ahead of the CME and had an appearance of a dome-like enhancement, and Cheng et al. [3], who investigated the formation and launch of a magnetic flux rope accompanying the launch of a CME.

5.2.3 *Summary*

The high quality images from *SDO* and the unprecedented capability of *STEREO* have enabled workers to continue to unlock more of the CME's secrets. These

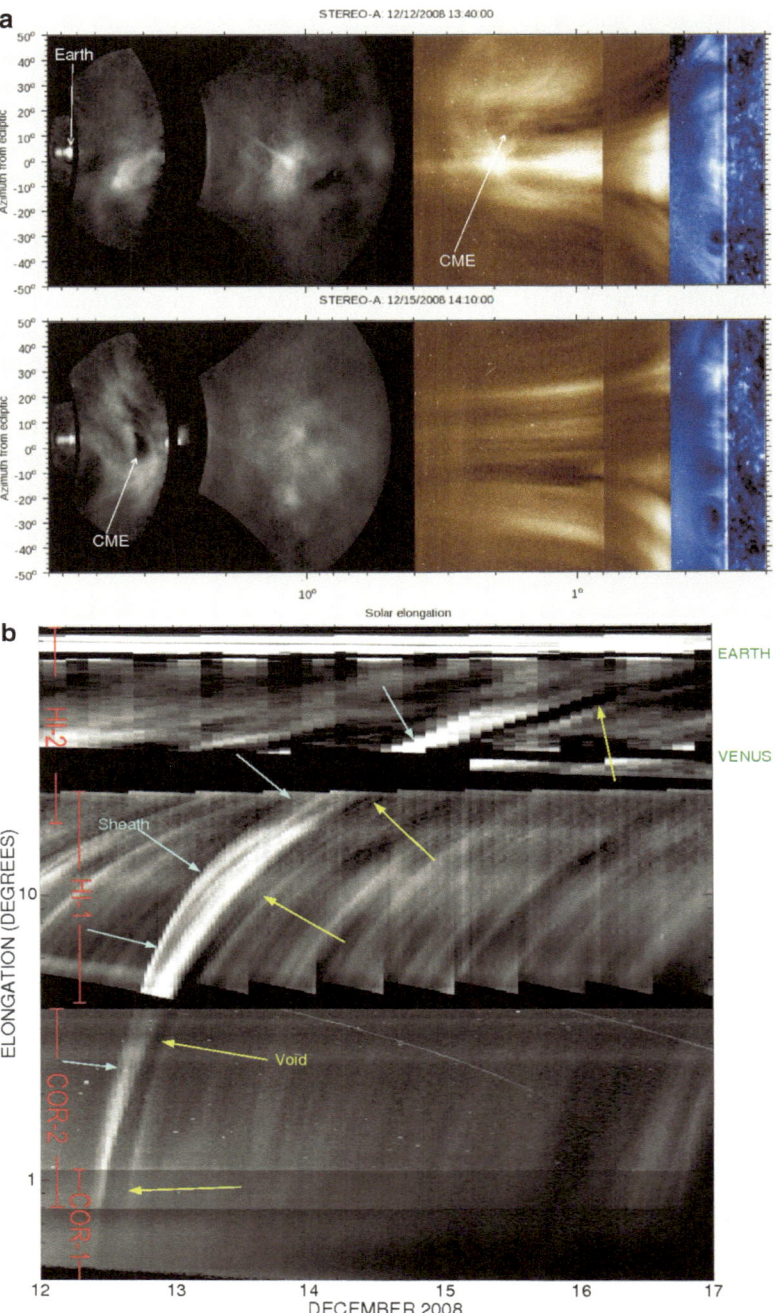

Fig. 5.4 (**a**) Two images of a CME that launched in December 2008 from the SECCHI pipeline [8, 20]. These are from *STEREO-A* and shows the fields of view of the entire SECCHI suite. The Sun is to the right (*blue*), and the fields of view from right to left are (indicated) COR1 and COR2

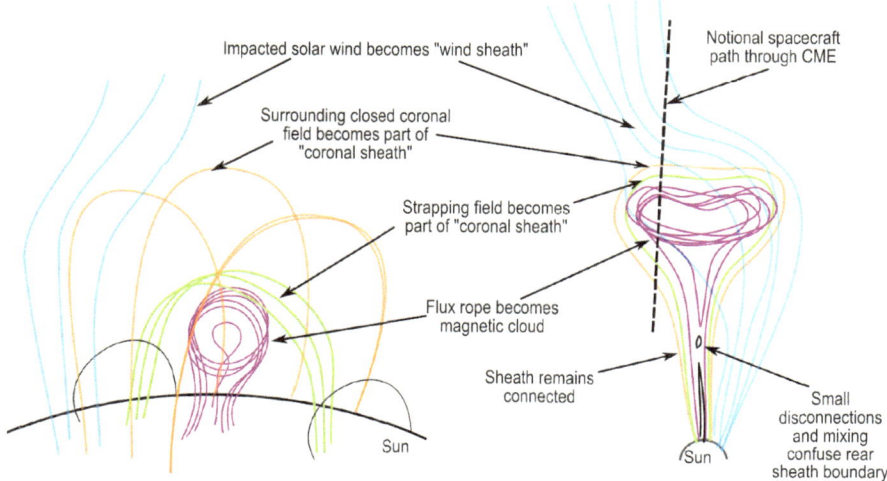

Fig. 5.5 Sketch of inferred magnetic topology for the CME of December 2008 [9]. Multiple flux ropes erupt from the corona, pulling the overlying strapping field with them. The coronal structure is preserved as it crosses the solar system, including plasma entrained in the strapping field that was originally in front of the CME flux rope

studies have focused on the primary questions that remain regarding the onset and evolution of CMEs. We are now almost in a position where we can observe the formation of the CME directly and measure its properties during the earliest stages of its launch. We can also track all of its components continuously through the solar wind, thereby improving our understanding of how these components evolve and interact. Finally, we can now compare our observations directly with in-situ measurements, enabling the combination of the detailed information provided by in-situ spacecraft with the larger picture produced by CME imagers.

5.3 Concluding Remarks

If you were to read any news article about the latest geomagnetic storm, the chances are that you will read that the storm was caused by a solar flare. This was most

Fig. 5.4 (continued) (*gold*) and HI-1 and HI-2 (*gray*). The Earth is shown in the field of view of HI-2 and is indicated. The CME (also indicated) is shown in the COR2 (*top*) and HI-2 (*bottom*) fields of view. The cavity (flux rope) of the CME is indicated in both images. The units in the x-axis are in elongation (angle from the Sun) and is log-scaled. (**b**) Elongation-time "J-map" of the same CME with elongation (log-scale) in the y-axis. The bright leading front (sheath) and dark void are indicated with the *arrows* [20]

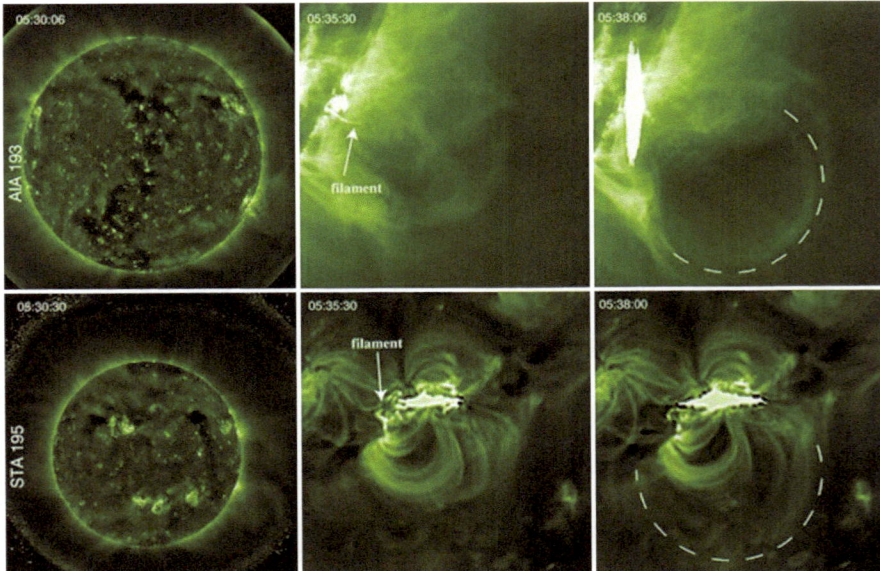

Fig. 5.6 *SDO*/AIA (*upper row*) and *STEREO*/EUVI-A (*lower row*) observations of an erupting active region, from Fig. 1 of Patsourakos et al. [38]. The *dashed lines* outline the expanding CME bubble and a filament, apparently associated with the initiation of the eruption, is marked by *arrows*

likely not the case. Most intense geomagnetic storms are caused by coronal mass ejections, which to this day continue to take a back seat to their more commonly known little cousins. The purpose of this Brief is to present an introduction to the coronal mass ejection. The importance of CMEs, not only to the evolution of the Sun, but also to space weather at the Earth, cannot be overstated. They are highly complex phenomena and are still misunderstood, despite decades of study. This book is mostly an abridged version of a more detailed book *Coronal Mass Ejections: An Introduction* by the author, but it has been updated to include the most recent and ongoing work studying CMEs. The most important message to take away from this Brief is the following: That it is the CME, and not the flare, that is responsible for the most significant space weather activity at Earth. CMEs also remove great quantities of energy and magnetic field complexity from the Sun, thereby providing a mechanism by which the Sun can shed this complexity. Research continues, but with each passing year we inch closer to a much more complete understanding of CMEs, their relationship with the Sun, and their impact on the Earth.

References

1. Byrne, J.P., Maloney, S.A., McAteer, R.T.J., Refojo, J.M., Gallagher, P.T.: Nat. Comm. **1**, 6, id. 74 (2010)
2. Carley, E.P., McAteer, R.T.J., Gallagher, P.T.: Astrophys. J. **752**, 36–43 (2012)
3. Cheng, X., Zhang, J., Liu, Y., Ding, M.D.: Astrophys. J. **732**, L25–L30 (2011)
4. Dasso, S., Mandrini, C.H., Schmeider, B., Cremades, H., Cid, C., Cerrato, Y., Saiz, E., Démoulin, P., Zhukov, A.N., Rodriguiez, L., Aran, A., Menvielle, M., Poedts, S.: J. Geophys. Res. **114**, A02109 (2009). doi:10.1029/2008JA013102
5. Davies, J.A., Harrison, R.A., Perry, C.H., Möstl, C., Lugaz, N., Rollett, T., Davis, C.J., Crothers, S.R., Temmer, M., Eyles, C.J., Savani, N.P.: Astrophys. J. **750**, 23–34 (2012)
6. de Koning, C.A., Pizzo, V.J.: Space Weather **9**, S03001 (2011). doi:10.1029/2010SW00595
7. de Koning, C.A., Pizzo, V.J., Biesecker, D.A.: Solar Phys. **256**, 167–181 (2009)
8. DeForest, C.E., Howard, T.A., Tappin, S.J.: Astrophys. J. **738**, 103 (2011)
9. DeForest, C.E., Howard, T.A., McComas, D.A.: Astrophys. J. **769**, 43 (2013)
10. Feng, L., Inhester, B., Wei, Y., Gan, W.Q., Zhang, T.L., Wang, M.Y.: Astrophys. J. **751**, 18–29 (2012)
11. Forsyth, R.J., Bothmer, V., Cid, C., Crooker, N.U., Horbury, T.S., Kecskemety, K., Klecker, B., Linker, J.A., Odstrcil, D., Reiner, M.J., Richardson, I.G., Rodriguez-Pacheco, J., Schmidt, J.M., Wimmer-Schweingruber, R.F.: Space Sci. Rev. **123**, 383–416 (2006)
12. Golub, L., Cirtain, J., DeLuca, E., Nystrom, G., Kankelborg, C., Klumpar, D., Longcope, D., Martens, P.: Proceedings of the AAS Meeting 37, Abstract 6.05, Durham, NH. Bull. Am. Astron. Soc. **38**, 226 (2006)
13. Gopalswamy, N., Yashiro, S., Kaiser, M.L., Howard, R.A., Bougeret, J.-L.: Astrophys. J. **548**, L91–L94 (2001)
14. Gopalswamy, N., Yashiro, S, Michalek, H., Kaiser, M.L., Howard, R.A., Reames, D.V.: Astrophys. J. **572**, L103–L107 (2002)
15. Gopalswamy, N., Yashiro, S, Kaiser, M.L., Howard, R.A.: Adv. Space Res. **32**, 2613–2618 (2003)
16. Harrison, R.A., Davies, J.A., Möstl, C., Liu, Y., Temmer, M., Bisi, M.M., Eastwood, J.P., de Koning, C.A., Nitta, N., Rollett, T., Farrugia, C.J., Forsyth, R.J., Jackson, B.V., Jensen, E.A., Kilpua, E.K.J., Odstrcil, D., Webb, D.F.: Astrophys. J. **750**, 45 (2012)
17. Hassler, D.M.: Proceedings of the AAS Meeting 40, Abstract 18.11. Bull. Am. Astron. Soc. **41**, 846 (2006)
18. Hassler, D.M., DeForest, C.E., McIntosh, S., Slater, D., Ayres, T., Thomas, R., Scheuhle, U., Michaelis, H., Mason, H.: Proceedings of the AAS Meeting 37, Abstract 37.06, Durham, NH (2006)
19. Howard, T.: Coronal Mass Ejections, An Introduction. Springer, New York (2011)
20. Howard, T.A., DeForest, C.E.: Astrophys. J. **746**, 64–71 (2012)
21. Howard, T.A., Tappin, S.J.: Solar Phys. **252**, 373–383 (2008)
22. Howard, T.A., DeForest, C.E., Reinard, A.A.: Astrophys. J. **754**, 102–111 (2012)
23. Liu, W., Nitta, N.V., Schrijver, C.J., Title, A.M., Tarbell, T.D.: Astrophys. J. **723**, L53–L59 (2010)
24. Liu, Y.D., Luhmann, J.G., Möstl, C., Martinez-Oliveros, J.C., Bale, S.D., Lin, R.P., Harrison, R.A., Temmer, M., Webb, D.F., Odstrcil, D.: Astrophys. J. **746**, L15 (2012).
25. Lugaz, N., Manchester, W.B., IV., Gombosi, T.I.: Astrophys. J. **634**, 651–662 (2005)
26. Lugaz, N., Hernandez-Charpak, J.N., Roussev, I.I., Davis, C.J., Vourlidas, A., Davies, J.A.: Astrophys. J. **715**, 493–499 (2010).
27. Lugaz, N., Farrugia, C.J., Davies, J.A., Möstl, C., Davis, C.J., Roussev, I.I., Temmer, M.: Astrophys. J. (2012, in press)
28. Ma, S., Raymond, J.C., Golub, L., Lin, J., Chen, H., Grigis, P., Testa, P., Long, D.: Astrophys. J. **738**, 160–169 (2011)

29. Manoharan, P.K., Gopalswamy, N., Yashiro, S., Lara, A., Michalek, G., Howard, R.A.: J. Geophys. Res. **109**, A06109 (2004). doi:10.1029/2003JA010300
30. Mierla, M., Davila, J., Thompson, W., Inhester, B., Srivastava, N., Kramar, M., St. Cyr, O.C., Stenborg, G., Howard, R.A.: Solar Phys. **252**, 385–396 (2008)
31. Mierla, M., Inhester, B., Antunes, A., Boursier, Y., Byrne, J.P., Colaninno, R., Davila, J., de Koning, C.A., Gallagher, P.T., Gissot, S., Howard, R.A., Howard, T.A., Kramar, M., Lamy, P., Liewer, P.C., Maloney, S., Marquè, C., McAteer, R.T.J., Moran, T., Rodriguez, L., Srivastava, N., St. Cyr, O.C., Stenborg, G., Temmer, M., Thernisien, A. Vourlidas, A., West, M.J., Wood, B.E., Zhukov, A.N.: Ann. Geophys. **28**, 203–215 (2010)
32. Mierla, M., Inhester, B., Rodriguiez, L., Gissot, S., Zhukov, A., Srivastava, N.: J. Atmos. Solar-Terr. Phys. **73**, 1166–1172 (2011)
33. Moore, R.L., Cirtain, J.W., West, E., Kobayashi, K., Robinson, B., Winebarger, A.R., Tarbell, T.D., de Pontieu, B., McIntosh, S.W.: Transactions AGU Fall Meeting, December 2010, Abstract SH11B-1655, San Francisco, CA (2010)
34. Möstl, C., Rollett, T., Lugaz, N., Farrugia, C.J., Davies, J.A., Temmer, M., Veronig, A.M., Harrison, R.A., Crothers, S., Luhmann, J.G., Galvin, A.B., Zhang, T.L., Baumjohann, W., Biernat, H.K.: Astrophys. J. **741**, 34 (2011)
35. Möstl, C., Farrugia, C.J., Kilpua, E.K.J., Jian, L.K., Liu, Y., Eastwood, J., Harrison, R.A., Webb, D.F., Temmer, M., Odstrcil, D., Davies, J.A., Rollett, T., Luhmann, J.G., Nitta, N., Mulligan, T., Jensen, E.A., Forsyth, R., Lavraud, B., de Koning, C.A., Veronig, A.M., Galvin, A.B., Zhang, T.L., Anderson, B.J.: Astrophys. J. (2012, in press)
36. Müller, D., Marsden, R.G., St. Cyr, O.C., Gilbert, H.R., The Solar Orbiter Team: Solar Phys. (2012, in press). doi:1007/s11207-012-0085-7
37. Patsourakos, S., Vourlidas, A., Kliem, B.: Astron. Astrophys. **522**, A100 (2010)
38. Patsourakos, S., Vourlidas, A., Stenborg, G.: Astrophys. J. **724**, L188–L193 (2010)
39. Rumberg, J.: The Solar Probe Webpage, Available via NASA/GSFC. http://solarprobe.gsfc. nasa.gov/ (2008). Cited 24 July 2008
40. Temmer, M., Veronig, A.M., Kontar, E.P., Krucker, S., Vršnak, B.: Astrophys. J. **712**, 1410–1420 (2010)
41. Temmer, M., Vršnak, B., Rollett, T., Bein, B., de Koning, C.A.., Liu, Y., Bosman, E., Davies, J.A., Möstl, C., Žic, T., Veronig, A.M., Bothmer, V., Harrison, R., Nitta, N., Bisi, M., Flor, O., Eastwood, J., Odstrcil, D., Forsyth, R.: Astrophys. J. **749**, 57 (2012)
42. Vandas, M., Odstrcil, D.: Astron. Astrophys. **415**, 755–761 (2004)
43. Wood, B.E., Wu, C.-C., Howard, R.A., Socker, D.G., Rouillard, A.P.: Astrophys. J. **729**, 70–79 (2011)
44. Wu, S.T., Wang, A.H., Gopalswamy, N.: Proceedings of the Magnetic Coupling of the Solar Atmosphere Euroconference and IAU Colloquium 188, pp. 227–230. ESA SP-505 (2002)
45. Zhang, J., Cheng, X., Ding, M.: Nat. Comm. **3**, 747 (2012). doi:10.1038/ncomms1753
46. Zhao, X.H., Feng, X.S., Xiang, C.Q., Liu, Y., Li, Z., Zhang, Y., Wu, S.T.: Astrophys. J. **714**, 1133–1141 (2010)

Glossary

Active region A region on the solar surface where the local magnetic field is concentrated. Sunspots and solar flares often occur within active regions.

Aly-Sturrock limit Energy limit showing that the completely open magnetic field state is more energetic than the closed state. This means that it is not energetically favourable for CMEs to simply erupt to an open field state.

Aphelion The furthest point to the focus of an elliptical orbit (point of furthest distance to a body an object is orbiting).

Bastille Day Event A famous and intensely studied CME/geomagnetic storm that erupted from the Sun on 14 July 2000.

(Plasma) Beta (β) The ratio of the plasma pressure to the magnetic pressure. So when $\beta < 1$ the magnetic pressure dominates the plasma, and when $\beta > 1$ the plasma pressure dominates.

Broadband Wide frequency range.

"Classic" three-part CME CMEs observed in coronagraphs are often described by a three-part structure. That is, a leading edge, followed by a turbulent low density cavity or sheath, followed by a bright filament.

Cone model A simple model that describes the CME as a spherical shell centered at the Sun.

Convection zone The region of the Sun where the plasma properties are such that energy is more efficiently transmitted through the medium via convection rather than radiation. This begins at around 0.6 $R_\odot$ from the center of the Sun and continues until near the solar surface, at 1.0 $R_\odot$.

Corona The outer atmosphere of the Sun that evolves into the solar wind. Sunward of the corona lies the chromosphere.

Coronagraph Device used to observe the solar corona. This is achieved by blocking out the bright photosphere of the Sun using an artificial disk called an occulting disk.

Coronal dimming A decrease in intensity in the solar corona often associated with the eruption of a CME. Coronal dimming is typically observed in EUV and X-ray but has also been detected in Hα.

T. Howard, *Space Weather and Coronal Mass Ejections*, SpringerBriefs in Astronomy, DOI 10.1007/978-1-4614-7975-8, © Timothy Howard 2014

Coronal hole Dark region in the solar corona corresponding to open magnetic field lines. Coronal holes are believed to be responsible for fast flowing solar wind streams.

Coronal mass ejection A large eruption of plasma and magnetic field from the Sun.

Coronal transient The original term for a coronal mass ejection.

Corotating interaction region A region of enhanced density in the solar wind brought about by an interaction between fast and slow solar wind regions, which corotate with the Sun. They are identified by similar in-situ signatures that are used to identify an CME, and so there is sometimes confusion between the two.

Cosmic rays Energetic particles typically with their origin outside the heliosphere.

Current sheet Electric current that is confined to a surface rather than a volume in space. They typically occur where there is no magnetic field but there are fields surrounding the sheet (i.e., a neutral line).

Cusp region The region in the magnetosphere where the boundary between the dayside closed magnetic field lines and the nightside open magnetic field lines occur. Through the cusp energetic particles from the solar wind are able to directly penetrate the magnetosphere to the ionosphere.

Dayside The hemisphere of the Earth facing the Sun.

Earth radius The radius of the Earth, $\sim$6,360 km.

Eclipse (solar) Where the moon passes between the Sun and the Earth, temporarily blocking out its light. This occurs because the apparent size of the moon from the Earth is the same as the apparent size of the Sun.

Ecliptic (plane) The plane in the heliosphere containing the orbit of the Earth. Most of the solar planets orbit in planes close to (i.e., within $20°$ of) the ecliptic.

EUV wave Wave traveling across the solar corona observed with EUV instruments. They are commonly associated with CMEs and CME/flare onset.

Elongation The angle between the Sun-observer line and the line from the observer to the point of interest. So, $0°$ elongation is the direction of the Sun, $90°$ elongation is the same plane as the observer, and $180°$ elongation is directly behind the observer relative to the Sun. Elongation can be regarded as an angular measurement of radial distance from the Sun, as images are projections and as such do not contain any depth information.

Filament A relatively dark region on the solar photosphere that appears as a dark line that varies in size and geometry. It is generally believed to be a concentration of relatively cool plasma suspended above the photosphere by magnetic fields. A disappearing filament often occurs when the magnetic structure erupts, carrying the cool plasma with it. Disappearing filaments are often associated with the eruption of a CME. When a filament is observed on the solar limb it is called a prominence.

Flux rope A column containing magnetic flux. Also known as a flux tube.

Geomagnetic field The Earth's magnetic field.

(Geo)magnetic storm A large disturbance in the Earth's magnetosphere and ionosphere. The major storms are typically caused by the arrival of a fast CME with a predominantly southward-directed magnetic field.

Halloween event A famous and intensely studied CME/geomagnetic storm that erupted from the Sun on 28 October 2003.

Halo CME A CME with a large component along the Sun-Earth line, and hence appears on projection to completely encircle the Sun.

Height-time plot Plot of distance from solar center against time, typically used to determine CME speeds. Measurements of the height of a CME are usually obtained from the leading edge of the structure as observed by a coronagraph.

Heliocentric Sun-centered.

Heliosphere The region within which the solar wind is contained, roughly a sphere from the Sun out to around 100 AU. For the purposes of this book the inner heliosphere is the region out to a couple of AU.

Heliospheric current sheet The surface where the polarity of the Sun's magnetic field changes. Generally described as a layer of dense plasma within which a strong current flows.

Helium abundance enhancement (enrichment) An enhancement of helium observed following interplanetary shocks. These were early indicators of CMEs.

Hermian Related to the planet Mercury.

Hinode Japanese spacecraft launched in 2006 designed to monitor solar activity with a suite of imagers.

Imager An instrument that captures images (camera).

In-situ Measurements made by instruments in direct contact with a phenomenon, in this case, when an CME passes through a spacecraft.

Interplanetary coronal mass ejection The interplanetary counterpart of a coronal mass ejection. Typically observed with in-situ spacecraft and often preceded by an interplanetary shock.

Interplanetary magnetic field The magnetic field that is carried away from the Sun by the solar wind. Every object in the solar system is embedded in both the solar wind and the interplanetary magnetic field. Because the solar wind rotates with the Sun, the interplanetary magnetic field rotates also, resembling an Archimedian spiral (the Parker spiral).

Interplanetary medium General term describing the medium containing the solar wind and interplanetary magnetic field. It is the medium within which the entire solar system exists.

Interplanetary scintillation The distortion of the signal from a distant radio source as a result of a dense structure passing between it and the observer.

Interplanetary shock A collisionless shock in the interplanetary medium typically brought about by the passage of a supersonic relatively dense structure such as a CME. Forward interplanetary shocks have an in-situ signature of a sudden increase in magnetic field, density and solar wind speed, while reverse shocks have a similar signature, except there is a sudden decrease in magnetic field.

Interplanetary transient The general term for a disturbance in the interplanetary medium. May be used as a more general description for an CME but does not exclusively describe them.

Ionosphere A relatively thin conducting layer of the Earth's atmosphere, immediately below the magnetosphere.

Jovian Related to the planet Jupiter.

The L1 Lagrange point The point between the Earth and the Sun where the gravitational effects of the Earth are exactly canceled by the Sun. This is located around 1.5×10^6 km from the Earth, or about 1 % of the distance between the Earth and the Sun.

Limb darkening A relative reduction of intensity on the surface of the Sun moving towards the limb, resulting from the curvature of the Sun and the nature of its radiation.

Line of sight The vector from the observer through the point of interest and out to infinity.

Magnetic cloud A magnetic flux rope typically associated with CMEs. It is often preceded by a shock and contains a highly twisted magnetic field structure. The in-situ signature of a magnetic cloud includes: (1) low temperatures, (2) strong magnetic field, (3) a smoothly rotating magnetic field vector. Magnetic clouds are also of long duration, lasting around 30 h on average.

Magnetic reconnection Where magnetic field lines from different regions connect with each other, such as where the interplanetary magnetic field interacts with the geomagnetic field.

Magnetic shear Region where the magnetic field runs almost parallel to its neutral line, so it is far from potential.

Magnetosphere Region of plasma enclosed by the Earth's magnetic field. It extends to around 10–15 R_E on the sunward side (dayside) and several hundred R_E on the anti-sunward side (nightside).

Magnetotail The nightside of the magnetosphere where the field lines are extended to large antisunward distances by the solar wind and IMF.

Mollweide projection A sky map equal-area projection where the latitude lines are parallel to the equator.

Narrowband Narrow frequency range.

Neutral line Region where the magnetic field is neutral, i.e., its consists of equal quantities of positive and negative flux. These may be regarded as magnetic "source" regions.

Nightside The hemisphere of the Earth facing away from the Sun.

Observer A general term describing an instrument or person looking at or measuring something.

Parker spiral General structure of the solar wind and interplanetary magnetic field. Field lines and plasma parcels move radially outward from the Sun but undergo corotation. This results in an Archimedian spiral structure.

Perihelion The closest point to the focus of an elliptical orbit (point of closest approach to a body an object is orbiting).

Photosphere The top layer of the convection zone of the Sun, where energy is transmitted to the surface via convection rather than radiation. This is the so-called "surface" of the Sun, and the brightest region observed in visible light.

Point P approximation A simple approximation for converting elongation to distance, when measuring CMEs. Point P assumes that the part of the CME being observed is spherical and centered at the Sun. This reduces the conversion to simple trigonometry: $p \sim \sin \varepsilon$, where p is the distance from the Sun in AU, and ε is the elongation.

Point spread function The response of an imaging instrument to a point source, or a function describing how a point source is spread across an image.

Polar caps Area mapping to the geomagnetic field lines that are open i.e., are connected to the IMF. Through these field lines energetic particles from the interplanetary medium may enter to lower altitudes of the Earth's atmosphere.

Post-eruptive arcade A region of hot plasma and highly-structured magnetic field from the low corona following the eruption of a CME.

Projection effects The effects of obtaining a two-dimensional image of a three-dimensional object. The image will represent a projection of the three-dimensional image into the plane of the sky relative to the observer.

Prominence A loop observed on the solar limb suspended above the solar surface by magnetic fields. Erupting prominences are often associated with the eruption of a CME. When a prominence is on the solar disk it is called a filament.

Ring current A current in the equatorial region of the Earth's magnetosphere brought about by the movement of trapped particles. It lies at a distance of 3–5 R_E from the Earth and circulates clockwise around the Earth when viewed from the north.

Running difference A sequence of images where each image has been subtracted away from the previous one. That is, in a running difference image $B_j = A_j - A_{j-1}$.

Sky plane The plane of the sky relative to the observer. Images of the Sun and heliosphere are projected into the sky plane.

Snow plow Where a CME accumulates solar wind material which cannot get out of the way. It is possible that much of the material observed by heliospheric imagers may be snow-plowed material.

Solar cycle The magnetic cycle of the Sun which lasts around 11 years. Throughout the cycle the magnetic complexity in the Sun increases, resulting in a large number of sunspots, solar flares and CMEs (solar maximum), and then the complexity decreases along with activity (solar minimum). At the start of the new cycle the magnetic poles of the Sun are reversed, meaning it takes two cycles to return to the original magnetic orientation.

Solar energetic particles High-energy particles originating from the Sun and observed in the heliosphere. Many are generally believed to be accelerated by solar flares and CMEs.

Solar flare A sudden increase in emission from a highly localized region of the Sun. Solar flares are generally broadband in nature, and can span the electromagnetic spectrum from below visible to above X-rays. They are often associated with CMEs, and are known to have effects on the Earth's magnetosphere and ionosphere.

Solar limb The edge of the solar disk, relative to an observer.

Solar maximum The maximum phase of the solar cycle: The period in the middle of the solar cycle where activity (e.g., sunspots, flares, CMEs) is maximized.

Solar minimum The minimum phase of the solar cycle: The period at the start and end of the solar cycle where activity is minimized.

Solar Radius The radius of the Sun, $\sim$695,500 km.

Solar surface The photosphere: The region on the solar disk where the Sun becomes opaque at the top of the convection region. This region is popularly observed with the Hα line.

Solar wind A body of plasma continuously-flowing away from the Sun. It may be regarded as an extension of the corona, and extends to around a 100 AU away from the Sun. It carries a magnetic field with it, known as the interplanetary magnetic field.

Solid angle The angle in three dimensional space subtended by an object at a given point. It is a reflection of the apparent size of an object to an observer at that point.

South Atlantic Anomaly A region in the south Atlantic ocean (just off the coast of Brazil), where the van Allen radiation belt makes its closest approach to the Earth's surface. Here, the radiation and energetic particle intensity is greatest.

Space weather A general term embracing many effects, including geomagnetic and magnetospheric activity. Large geomagnetic storms are types of severe space weather at the Earth.

Sudden (storm) commencement A sudden increase in geomagnetic activity, usually triggered by the arrival of an interplanetary shock.

Sudden ionospheric disturbance Sudden increase in ionospheric density in the D region of the ionosphere, indicative of an increase in geomagnetic activity.

Sunspot A dark region on the Sun indicative of solar activity there. Sunspots arise from emerging magnetic fields from below the photosphere.

Supercritical twist A highly-twisted flux rope state which may enable the launch of a CME.

Termination shock The region of the heliosphere where the solar wind slows down to sub-sonic speed. This occurs at distances around 80 AU.

Thomson scattering The scattering of electromagnetic radiation from a charged particle, brought about by the acceleration of the particle by incident radiation.

Thomson surface The resulting sphere from obtaining the locus of all points where any line of sight is perpendicular to the solar radial vector. For an observer at the Earth, the Thomson surface is a sphere with a diameter of the Sun-Earth line and the surface crossing both the Earth and Sun.

Triangulation The technique by which the three dimensional location of a point can be determined when observed from multiple locations. The technique involves the application of geometry.

Type I burst Short-duration, narrowband radio bursts occurring during storm periods.

Type II burst Long-duration, varying frequency radio bursts driven by CME shock acceleration.

Type III burst Short-duration, broadband, varying-frequency radio bursts driven by solar flares.

Type IV burst Long-duration radio bursts that often follow Type II bursts.

Type V burst Radio bursts that often follow Type III bursts.

Van Allen belts A region of energetic charged particles around the Earth, trapped by the geomagnetic field.

Index

T. Howard, *Space Weather and Coronal Mass Ejections*, SpringerBriefs in Astronomy,
DOI 10.1007/978-1-4614-7975-8, © Timothy Howard 2014